It's Different in the Country

LIZ POTTER

CORGI BOOKS

To Brian, who called my bluff by giving me a typewriter

IT'S DIFFERENT IN THE COUNTRY

A CORGI BOOK 0 552 12681 0

First publication in Great Britain

PRINTING HISTORY
Corgi edition published 1985

Copyright © Liz Potter 1985

This book is set in 10/11pt Plantin

Corgi Books are published by Transworld Publishers Ltd., Century House, 61–63 Uxbridge Road, Ealing, London W5 5SA, in Australia by Transworld Publishers (Aust.) Pty. Ltd., 26 Harley Crescent, Condell Park, NSW 2200, and New Zealand by Transworld Publishers (N.Z.) Ltd., Cnr. Moselle and Waipareira Avenues, Henderson, Auckland.

Made and printed in Great Britain by
Hunt Barnard Printing Ltd., Aylesbury, Bucks.

Chapter 1
A House Cow

'Yuck,' grumbled Ted, peering into his mug of coffee where I could see a good half-inch of sludge, murky in colour and gooey in consistency. I knew what the trouble was, I'd forgotten to buy fresh milk once again and had used powdered in the coffee.

'I'm sorry,' I said, trying to sound apologetic but as it was the third time in as many weeks I wasn't expecting love and understanding.

'This is horrible,' continued Ted, his blue eyes looking hurt and hard-done-by. 'In any case it's ridiculous to be having this stuff when we're living in the country.'

'It's because we're in the country that I don't go shopping very often so then I have a lot to get and I forget more, so – er – well . . .' I trailed off realizing how foolish I was sounding. But in my defence there was some truth in it. Our cottage is in the middle of Bodmin Moor, the nearest town is six miles away and the village is three miles off. To be efficient one makes a weekly list, has a day's shopping and all's well. In my case I make the list and leave it on the table.

'I've been thinking,' Ted tried a spoonful of sludge on the dog. To his satisfaction the dog turned away from it in disgust. 'I think we should have a house cow, we've three fields and a barn all unused, and if we buy one in calf we can have her calf and get another one from market so that she suckles two and we'll have milk and profit when we sell them.' He sat back in his chair and awaited approbation.

'We've got to buy it first.' I said a bit sharply. 'We haven't any money.'

'Not for frivolous things,' agreed Ted, well aware that I was still smarting from his refusal to let me buy a feather-filled continental quilt on the grounds that he wanted to sleep at night, not hatch out. 'But a cow would be an investment. Jersey cows are pretty,' he added as an incentive.

'Well,' I began.

'Anyway, I can milk her,' I was reassured as he struggled into his overalls prior to working outside. Then as he went through the door he said, 'It won't take you long to learn to milk, it's easy!'

My heart did a hiccup. I loved the country and I was happy to be a part of it, but Ted had been in farming on and off most of his life. I was still bemused at the whole situation I found myself in. The children Steve and Kate, if you could call them children at the ripe old ages of twenty-one and nineteen years respectively, had flown the nest and had flats in the city in order to get to work more easily and cheaply, but they returned home at weekends with plastic bags of washing and insatiable appetites. They listened with thinly-veiled puzzlement to our plans to create a pig farm from scratch, exchanged glances with each other and decided that senile mental decay had set in early. The cottage we'd moved into was small but it boasted three fields and a decayed barn which Ted was even now rebuilding. Until we created pig houses and filled them we had no income, and the precious bit of capital we did have was earmarked for the stock. Meantime, we lived as meagrely as we could. I was facing up to rural living at close quarters, starting with a cow's quarters it seemed. For the next couple of weeks I excelled myself remembering to buy milk and I thought Ted had forgotten his idea. Not so.

'Here we are,' announced Ted triumphantly one evening from behind the local paper. 'Just the thing – Jersey house cow due to calve in two weeks time, very friendly. The price is OK too, I'll ring up now and we can go and see her tomorrow morning.'

I dressed suitably for the occasion: thick anorak, jeans and wellies. We hadn't far to go, a few miles of narrow twisting lanes that took us to the other side of the moor. An elderly man appeared round the corner of a stone barn. He acknowledged us by jerking his chin in our direction.

'About the cow?' Ted wasted no words either.

'Thass 'er,' the farmer jerked his chin toward a field gate. Through the gusts of rain that blew horizontally across the moorland landscape I could see a cow sensibly under an umbrella of a hawthorn tree. The gate was, of course, fastened with baler twine. Even I knew enough not to try and undo someone else's baler twine – he who knots can unknot – that's what I say. Once the string was unravelled and Ted helped to lift the gate open we walked across the field to inspect the animal. The farmer stood silent with his hands in his pockets while Ted walked round the cow. She stood quietly as Ted bent double inspecting her undercarriage and feet. All I could do was assure myself that she had four of some things and two of others. What I was pleased about was the fact that she let me rub her head.

The rain began to penetrate my anorak and I wondered to myself how Kate would react to a house cow at the weekend.

In fact, she was pretty scathing when I told her, reminding me that we had three cats and two dogs already and insinuating that at my advanced age that should be sufficient.

'Transport?' queried the farmer.

'Can you?' asked Ted.

'Where be you living?'

The transaction was completed and we arranged for the cow to be brought to us that evening with the minimum of words. As we got into the car to leave the farmer came up to my side. I looked expectantly at him, was he going to speak?

'Er's called Connie,' he informed me and jerked his chin at me before turning away.

'Not much of the high pressure salesman, was he?' I said.

'No need. If I hadn't intended buying I wouldn't have gone there and the price was in the paper. I offered him twenty pounds less, he accepted and he's going to deliver her free. He's got what he expected and so have I.'

'Did you really do all that?' I was amazed, I'd only heard a few words exchanged and hadn't caught all of them.

'You were there,' Ted pointed out.

'I'm not used to this sort of thing yet,' I said and made a mental note to listen more carefully in future. I'd missed most of that deal because I hadn't realized it was taking place until it was over.

Connie arrived in a large lorry just as it was getting dark. We had to unload her on our new gravel drive. She came down the ramp in dignified style. With the briefest pause she looked around at us lit up by the outside light, then in a brisk manner she scrunched across the gravel into the darkness beyond. All was silent as she walked on to the grass of the field. We knew she couldn't go far so we left her to settle in.

In the morning we got up early to see her, only we couldn't because she was gone. Around our fields are old stone walls that have crumbled in places and eventually we found her on the other side, having discovered a nice boggy patch instead of the green field we'd intended for her. It didn't need too much persuasion to get her back, but the rest of the day was spent in fencing her in and organizing water for her. There were no troughs in any of our fields, although the Water Board were due to install some as soon as they had acquired some of what had been part of the farm before we arrived. Where troughs had been was no longer our land but would soon be part of a new reservoir which even now was being created; meanwhile we had to provide water for Connie. Ted found a galvanized bath and we filled it with water in buckets laboriously collected from the house. Water is heavy and it wasn't until my third journey that I made a discovery. The bath leaked. It was then I had a brainwave. I remembered a plastic cover that had been on a new mattress and I lined the bath with it. Ted was quite impressed and Connie ambled up most interestedly as we refilled the bath, she emptied it as fast as a kid with a Coke. That's when Ted had a brainwave to match my own.

'Tell you what,' he said, 'We'll use the milk churn. I'll fill it with water, hump it on to the wheelbarrow and wheel it along the road and tip it over the fence into the bath.'

'Er, What milk churn?'

'The one in the kitchen, of course, we've only got one.'

'D'you mean the kitchen stool?' I was rather proud of the stool. I'd found the old churn outside and I'd painted it up and covered the lid with some foam and a piece of material on top, which made a serviceable if somewhat heavy stool. So – there was Ted pushing the wheelbarrow with me steadying the churn, making a wobbly progress along the road to Connie's field when a car of holiday-makers came along. They slowed down and looked extremely puzzled. I've often wondered what they made of that gaily-coloured churn being pushed along the lane.

Connie was grateful for her water and after much serious drinking we left her rubbing her chin on the edge of the bath.

Steve and Kate arrived the following weekend in Steve's car but with Kate at the wheel. It says a lot for Kate's persuasive ways. As they came in it appeared that her brother was not impressed with the driving and that he felt she should take up flying where she would find more space.

'She's not really that bad is she?' I asked in my Worried Mother's voice.

'She won't take it seriously,' complained Steve. 'Last time I let her drive it had been raining and she was zig-zagging all over the road trying to avoid the frogs and singing at the top of her voice 'A frog he would a-wooing go and—'

'It's awful to think of them going pop when you run over them isn't it Mum, you always zig-zag don't you?'

'Well, er, yes I do. Poor little things.'

Steve looked at me pityingly.

'We've got our house cow,' I announced in order to change the subject. They were both now looking at me pityingly. Well, perhaps driving like an idiot to save frogs was overdoing it a bit, but I felt reason was on my side about having a house cow. 'Come and be introduced,' I said firmly. They followed me politely, no doubt exchanging glances behind my back.

Connie was placidly sitting and chewing cud. She watched our approach calmly and made no effort to rise which showed she felt at home. I'm not sure what I expected from my offspring but they were quite kind in a patronizing sort of way. Kate admired Connie's beautiful eyes which were indeed attractive, having dark markings around them, an effect I'd tried on myself once or twice and ended up looking as though I'd had a sleepless night whereas Connie looked like Cleopatra. Steve remarked that he thought she seemed a bit fat and I scathingly pointed out that it was hardly surprising seeing that she was due to calve in about a week's time. That bit of information impressed them and I took advantage by telling them that we'd soon be having home-grown milk in our tea.

'And who's doing the milking?' enquired Steve loftily.

'Well, I'll soon learn,' I snapped, crossing my fingers behind my back. 'You could both learn, too.'

A look not far removed from sheer horror flitted between them, and they began to wander back to the house. I bent down to Connie and she let me rub between her ears. I smoothed her neck and finally let my hand travel along her body. The movement made the calf inside her move. Connie and I looked each other in the eye and I felt we had an understanding. We both knew a bit about motherhood. I twiddled her ears and said a few encouraging words before turning toward the house. Steve and Kate were watching. I saw them shrug their shoulders.

As it happened, Connie calved just four days later, in the night. Ted and I had taken to a last stroll at night to see her. It was a very pleasant chore to wander to her field and usually the dogs were around as well, although Connie didn't encourage them too near but she didn't mind Buttons the black cat at all, he was allowed to rub himself against her legs. This particular evening Connie was standing near a small hawthorn tree, not lying as she usually was by this time of night. Ted was sure she would calve before the morning. He set the alarm clock for 1 A.M. I heard the alarm all right but I couldn't wake up. However Ted did but he was back quite quickly.

'Nothing doing yet,' Ted snuggled up to my back, he was cold and I shot across the bed to get away from his icy knees.

'Didn't you put on any clothes to go out?' I squeaked.

'I had on my pyjamas,' Ted defended himself. 'It's not cold out.'

'You could have fooled me,' I grumbled. I decided to do the noble thing and go and look for myself the next time the alarm went off. It was just as well because Ted didn't move a muscle. I pulled on a few garments and went to the field. It was just getting light and I could see a smudge of mist here and there on the moor beyond the fields. I think I woke up a couple of birds, otherwise all was still. Connie was in the corner. I walked up to her and rubbed her head. I went round to the business end and saw she was holding out her tail, then she started sort of paddling with her feet and gave a low moo. I stayed with her a while and nothing more happened. I started to shiver and went back to the house. The dog woke up and welcomed me back in as he would any burglar. I was wide awake by now so I made a cup of tea. I went out to Connie again and this time I could see the tip of a hoof protruding from under her tail. I raced back in to get Ted. Together we advanced on Connie but as we got nearer we saw her give birth to a very limp calf that slithered on to the ground and didn't move. Ted did everything he could but there was no life in that calf, it was perfectly formed and a good size, just dead. Connie turned to study it after Ted had finished. She breathed heavily over it, lowed a couple of times and then quietly walked away. She accepted the situation better than I did. However, practical things had to be done.

'We'll go to market this morning and get a calf to put on her, she's brimming with milk,' Ted decided.

There were pens of pathetic calves in the market and normally I avoid looking because I want to save them all. This morning was my big chance to take one to a good home. Ted and I stood around the pens with other farmers studying the little animals.

'We want one that matches,' I began. Ted took my arm in a vice-like grip and whisked me away.

'For goodness sake,' he began. 'If you're going to talk a load of cr— rubbish like that don't stand anywhere near me. We want a good calf, not a matching accessory.' And he stalked back, leaving me. I felt hurt, I was sure the right colour would matter a bit.

'Don't worry, Missus, they don't notice the colour and I expect your man'll rub the afterbirth over it anyway if he's trying to get a cow to take a strange calf,' said an older man who had overheard Ted and me. I smiled wanly at him, realizing that I'd made a bit of a fool of myself. I walked to the market café and got a cup of tea. The trouble was I wasn't experienced in these farming matters, enthusiastic, yes, but still wet behind the ears. I remembered that the best way to get a cow to accept another calf was to rub her afterbirth over it and then present it to her, and if you did that the colour wouldn't show in any case. I cooled down and prepared to forgive Ted, I must have embarrassed him horribly. The café door opened and there was Ted looking like a large Noddy in his blue overalls. With his favourite woolly hat on and one bit of fair hair showing and a satisfied smile he looked every inch the farmer. I'm married to a farmer, I reminded myself, I've got to become a farmer's wife.

'I'm sorry, Ted,' I began.

'No, it's me that should be sorry, luv, it didn't matter. It's been a long time since I was at a market to buy and I was worried in case I made a fool of myself by buying a duff animal but I reckon you don't forget what to look for. I've got two, and one matches,' he said with a grin.

'Oh Ted, I am glad. I'm going to try to be a practical farmer's wife.'

'You're not the shape, you'll have to put on weight,' he answered. 'Look, they're mostly twice as a big as you, well, around anyway.'

It seemed true, most farming ladies are buxom, perhaps it's the food, or the healthy living or just plain contentment. At five foot ten inches, if I put on weight I'll be huge I thought, although I wistfully wondered if I'd get a bust if I increased my weight, it might be worth trying. Ted interrupted my dreams.

'Come on then, let's take our new babies home.'

I rode in the back of our car with the calves. Ted had folded the back seats down and I sat on the floor with an arm around each calf to prevent it from falling. They looked so pathetic, large-eyed, moist-nosed and knobbly-kneed. Their fur was so soft and they were so bewildered that I wanted to hug them to reassure them. However, one squirted a yellowish deposit on to me and the other sucked the collar of my anorak so I was rather relieved to hand them over to Connie. She was suspicious to say the least. Ted had kept her afterbirth and he rubbed it generously over each calf. They looked grubby, their coats all matted, and they smelt awful. Connie agreed with me. It took Ted a long time and a lot of patience finally to get both calves sucking but the satisfaction was great. It only took another day before Connie had accepted them and was to be seen happily licking and feeding them. She found the sunniest corner for them to sleep in and she guarded them from the old dog who wasn't the least bit interested in calves unless they were in his favourite watering-spot.

Once the calves were established Ted decided to get Connie used to being milked. The idea was to have just enough for us and the rest for the calves. Connie was a good milker and it didn't seem long before we were getting more than enough, so I decided to try my hand at butter. I had a recipe but the churning is the challenge. I stood half a bucket of milk overnight and topped it up with some more from the next morning, and left it a couple of hours. I poured off the top half from the bucket and that was when Ted had a stroke of brilliance. He got his electric drill and fixed a wooden spoon where the drill goes, he switched it on to slow speed and I held it in the bucket of creamy milk.

Cleaning milk splashes off the walls and ceiling is not fun so the next time I covered the bucket with a teatowel first, held up the edge and very gingerly placed the drill under cover into the bucket. After about five minutes the butter 'came', that is, solid nodules of yellow formed which stuck together in increasingly larger lumps. That was one of the most satisfying moments of my life, strangely it proved to be a bit of beginner's luck as it never 'came' so quickly again. I thoroughly enjoyed squeezing the butter milk out and kneading the salt in but the best part was shaping the butter into pats and placing them on dishes ready for use.

So far, you will notice, I've avoided the subject of milking. Of course it had to happen, and one fine morning I was escorted into the shed by Ted in one of his educating moods. To start with, there was nowhere to tie up a cow.

'No need to tie her up, she's very good. Just give her some dairy nuts in a bowl and off you go.' Ted proceeded to show me how efficiently he and Connie got on with milking; she stood still munching happily and Ted did the bell-ringing bit. Easy. Now I had a try. Connie kept fidgeting and moving her bowl forward so that she had to follow it which meant that I had to follow her. This was not easy because Ted used a log of wood to sit on while he milked and it was difficult to shift. Sitting on a rather wobbly lump of wood I attempted to get milk out of Connie's teats. They felt like dried-up sausages after they'd been left unwrapped in the fridge, except Connie was warm. I buried my face in her hairy side and thought how nice a cow smells.

'Come on then,' said Ted. 'I can't stand here all day waiting. Get started.'

I got started, I grasped a teat in each hand and squeezed. Nothing happened. I tried again. A minute bead of milk formed on the end of one teat and dropped off on the floor. I tried once more and got a small squirt which went up my sleeve. Ted laughed as he leaned against the wall.

'You're getting the hang of it,' he said. 'Hold the teat near the top and gently squeeze it out of the end, but don't pull.'

I did manage to get a few decent squirts but couldn't seem to direct them so some went in my wellington boot, a little more up my sleeve and very little in the bucket. I was so slow. When I had progressed to milking Connie all by myself I took the radio with me and heard most of Jimmy Young before I'd finished. My back used to ache, my fingers went numb, my neck ached and I felt as if I'd never straighten up after a milking session. But it was satisfying. As I got quicker it became one of my favourite jobs. I used to marvel at the many hours people sat milking in the not so distant past when there were no milking machines. I was glad Connie was our only milker.

I got fond of Connie, told her my troubles while I milked her, rubbed her head and sat with her in the fields sometimes when the weather was warm enough to read a book out of doors, and used her as a backrest which she never minded. I noticed sometimes when she got up she creaked in the joints a bit, and sometimes she walked a bit stiffly but she was well enough in herself. Even when the calves were weaned completely she didn't put on any weight but she ate and seemed content.

In the late autumn we started letting Connie come into the barn that Ted had finished. We had some hay in there but partitioned it off so that Connie and two, now large, calves couldn't pull at it. One evening was cold and a gale was blowing up with chilly rain. Ted made sure the three 'stock' were in the barn with the door shut so that it couldn't get blown off its hinges.

'I'm glad they've got that barn to go in,' I said that night as Ted and I sat in front of the fire listening to the wind and rain.

'We're spoiling them really, they'd be all right outside, like the moorland cows.'

'Well, Connie's getting on a bit.'

'Yes, I was looking at her the other day, I think she's getting on quite a lot.'

'She seems happy enough, doesn't she?'

'She should be with you giving her nuts every five minutes, and three fields to graze.'

In the morning Ted went to let Connie out of the barn and he came into the kitchen a moment later.

'Connie's dead,' he announced.

'She's dead? Oh, no – she wasn't ill, was she, and we didn't know?'

'I don't think so, I can't be sure but it looks natural, old age I think it could be.'

'Oh Ted, I'd got so fond of her, what happens to her now?'

'I'll phone the knackers,' he said sounding angry and I knew he was sad too, Ted often sounds angry when he's sad and doesn't want to show it. I didn't want to be around when Connie was taken away so I went for a walk with the dog on the moor. I didn't cry, I knew animals couldn't live for ever and Connie had died a happy cow. The thing I hated most was the cheque that came a week later, it said 'Hide of cow – £8 – it was from the knackers.

Chapter 2

Enter a Goat

I missed Connie and to be quite practical, I missed her milk but I hadn't the heart to consider another cow. Ted mentioned it once or twice but we didn't get around to doing anything about it. Connie's two calves were grown up and Ted was busy planning our future pig houses and building up broken-down walls and outbuildings. He appeared to glean much helpful information from the locals in the Fox and Goose, some of which I listened to gratefully and some I discarded as bar-room chat. One morning over a cup of coffee Ted made a remark about goats. I knew nothing of goats and shared a rather general opinion that people who kept them were somewhat eccentric.

'I think we could have a goat.'

'Could we?'

'A goat would provide milk but not as much as a cow does.'

'Isn't their milk a bit funny?' I asked to be polite because I didn't seriously think we were likely to have one. I mean, after all, a goat?

'No, I don't think so, not if it's treated right. You see, I know where we could get one cheap.'

'Oh yes?' I became more alert at that, Ted couldn't resist a bargain whether it was something he wanted or not.

'Yes,' Ted sounded over-casual. 'There was this chap in the Fox and Goose the other night, he has a goat and he desperately needs a good home for it. Matter of fact, I think he'd give it to us if we wanted it.'

'Do we want it?' I enquired anxiously. 'Anyway, what's wrong with it?'

'Nothing, nothing at all.' Ted sounded hurt. 'Just because I heard about it in the pub you think there's bound to be something wrong.'

'That's right,' I agreed. 'It's where you met the chap who was going to come and repair the Rayburn the next day and he never turned up, not to mentioned the one who promised us firewood and what about the one who—'

'All right, I know, but this is different. This chap is moving house and will only have a garage to keep the goat in. You need a proper place to keep a goat and he won't have one.'

'We haven't got one.'

'Yes we have, I can easily partition off a space in the big barn.'

'But I don't know the first thing about goats,' I wailed.

'This one would be just right for you to start on, it's only a baby.'

'Have you agreed to have it?' I demanded.

'Not without asking you first,' he answered righteously.

'And you're asking me now?'

'Well, yes. I haven't promised anything but I said I'd ask. You see, he's having to move sooner than he thought and if we could have it for a while, just until he finds another home for it?' Ted was pleading now.

'You mean if we take it we're not committed in any way?'

'No, not at all.'

'Well, I'll think about it.' I hoped that perhaps in the meanwhile this chap would lumber someone else with his goat. 'When are you likely to see him again?'

'He, er, well he might pop in today, that is, if he's passing.'

I felt pretty sure that he would be passing so I wasn't standing on one leg in amazement when a car came into the yard about half an hour later. What did surprise me was the sight of a small white goat sitting up in the back seat.

'You must be Liz,' said the rather good-looking man who got out and gave me a dashing grin. 'It's awfully good of you,' he continued.

I wondered what was awfully good of me but I had a pretty shrewd idea. Ted could be terribly goodhearted after a pint or two.

'This is Sally,' the man introduced his wife who looked friendly and also gave me a lovely smile. Then Ted came round the corner also smiling. I felt it was a foregone conclusion but determined to make the most of it.

'You've met John and Sally, I see,' he said heartily.

'Oh yes,' I agreed. 'We've met.'

'I do hope you don't mind but we've brought Topper with us. We thought it would be easier if you actually saw her,' said Sally. Easier to what, I wondered.

'Shall I let her out?' asked John. 'She's due for her feed and we brought it with us.'

'You're longing to meet her, aren't you?' Ted said.

'Oh definitely,' I murmured.

John opened the car door and Topper made a hurried exit. I had to admit that she was a delightful little thing. Pure white, dainty feet, and the prettiest face with bright amber eyes. I noticed at once the bar-shaped pupil which gave her a pixie-like expression. She trotted up to us in a self-assured manner.

'Maaaa-aaa-aaa,' she said and looked at us expectantly. Sally produced a lemonade bottle of milk with a teat on it.

'I expect you'd like to feed her,' she stated and thrust the bottle in my hand. I couldn't refuse because Topper attached herself to the teat almost simultaneously. She sucked at it with the rhythm of a jungle beat. She closed her eyes in ecstasy and emptied the bottle in next to no time. She released the teat and there was a hiss of air as it rushed into the vacuum within. Topper stood back and regarded me with a questioning air. I felt everyone's eyes upon me. This was the time to explain that I knew nothing of goats, that we had decided not to have any animals until we'd sorted out the pig buildings and that if we were to have a goat it would be to give us milk and not the other way around.

'She's sweet,' I heard myself saying. 'Perhaps we could have her—' I was going to say that perhaps we could have her until somewhere else could be found but Ted interrupted.

'I knew you'd like her,' he said smugly. 'I told you, John, didn't I?'

'That's right,' agreed John. 'You were quite sure it would be all right.'

'But I thought—' I began.

'Could we possibly leave her with you today do you think?' This was John with a winning smile.

'You see,' went on Sally, 'we're moving sooner than we thought.'

'That'll be all right won't it?' Ted was enthusiastic and staring at me.

'All right,' I answered meekly. I knew when I was beaten.

'Maaaa-aaaa-aaaa,' said Topper and started to chew the hem of my skirt.

'Why don't we all go indoors and have a cup of coffee?' Ted suggested, avoiding my narrow-eyed look which I'm always trying out but which has no effect on him whatsoever.

'We haven't any milk I'm afraid, I haven't been shopping today.' I felt I had to make a point.

'That's all right,' said John quickly. 'Sal brought bottles of milk for you for Topper.'

'That's right, I'll fetch them and the bag of feed. You know, in case you did say she could stay.'

'Brought her bed as well?' I knew I was being sarcastic but Sally looked crestfallen.

'I never gave it a thought,' she said sadly. 'Topper sleeps

on straw but she has a sack she's fond of, I'll drop it in.'

'Never mind,' I grinned. 'I was only being funny. We've got some straw around and Ted's going to make Topper her own little bedroom this afternoon. Aren't you Ted?'

'Oh yes. Yes of course,' agreed Ted reluctantly. I knew he'd set aside an hour for watching a rugby match on television but I felt he didn't deserve it entirely.

'Can we leave Topper somewhere just for now and I'll put the kettle on.'

'She can stay in the back room,' decided Ted. 'You all go on in and I'll shut the door. She can't come to any harm there for a bit.'

Our outside room is the area between the door to the yard and the door in to the kitchen. It is more of a covered passageway. It is also a glory-hole for the washing machine, deep freeze and a sink. Scattered untidily about are oilskins, wellington boots (five left and four right at the last count) bits of things that Ted fondly imagines will be useful one day and a cupboard with empty bottle for when we have time to make wine. The dogs are supposed to dry off there before they come in but they usually get in front of the fire while I struggle with my wellies. John gave Topper a small bowl of feed and we left her.

Over coffee we discovered that John and Sally made home wines and that they weren't moving far away so we made plans to see them. They seemed very nice people and I'd just forgiven them and Ted for press-ganging me into being a goat owner when a long pitiful bleat was heard.

'I wonder what she's up to?' muttered John looking a trifle apprehensive.

'I'll have a look,' said Ted and opened the kitchen door. 'She's not there!' he exclaimed.

'She's bound to be,' I said. 'Nowhere she can go, unless you didn't shut the door.'

'The door's still shut, where on earth . . . ?'

We all crowded to the doorway and another bleat was

heard. Our eyes moved up and there on top of the cupboard was Topper. There was about two foot of clear space between the top of the cupboard and the ceiling and Topper was sort of cowering there.

'Good Lord' I exclaimed. 'How did she get up there?'

We worked it out that she must have jumped on to the freezer and then up on the cupboard, but why she did it defeated us. It wasn't easy to lift her down because in spite of the pathetic bleats I don't think she wanted to get down. She wanted to be noticed and she was, so she was happy to stay up there. Once on floor level Ted firmly guided her into the barn.

'Sorry about that,' John said.

'Fancy her getting up there,' added Sally brightly.

'Just does these things for attention. Always up to something,' John continued. 'The other day she—'

'We really must be going,' interrupted Sally. 'I've loads of shopping to do.'

'Have you? I thought you'd done it all yesterday,' grumbled John.

'Goodness me no, there's lots more I want.' Sally made her way to the door and I followed her out to the car. They were both most effusive in thanking me for taking Topper and they assured me that they didn't want any payment for her. To be honest it hadn't entered my mind so I felt and bit mean and thanked them in return. I was waving goodbye when I heard another bleat from the barn. Poor little goat, I thought, I bet she's unhappy and wondering where she it. I went in to find that she had cornered Buttons the black cat against the hay bales. His fur was up and his golden eyes glowed indignantly. I scooped him up and Topper trotted hopefully behind us to the door.

'You only have milk three times a day,' I told her.

'Maaaaa-aaaa-aaaa,' she answered demandingly.

'What shall I do with her?' I asked Ted.

'Pop her in the barn, she'll get used to a new home in a day or two.'

I couldn't bring myself to do that, so I shut the gates and let her stay in the yard where she could see me through the kitchen window and I could keep reassuring her. She nibbled daintily at the bits of grass growing at the base of the buildings and finally sat on the back door mat and went to sleep.

As I soon discovered it was impossible to leave Topper anywhere, she was usually there first, wherever the action was, and that included callers. The first time Steve came after we had Topper he left his car, the pride of his life, in the yard as usual and he and I sat chatting in the kitchen.

'Seen Topper, have you?' enquired Ted when he came in.

'Yes, indeed, the minute I arrived,' answered Steve feelingly. He had suffered a close encounter of the unexpected kind as soon as he got out of the car.

'I meant, have you seen what she's doing now?'

By the tone of Ted's voice I guessed Topper was up to no good and I hastily went to the window. There she was on Steve's car, prettily balanced on the bonnet and nibbling gently at the windscreen wipers. Steve shot out and shouted at her. She merely raised her head enquiringly and gazed steadily as he waved his arms like windmills. When he actually lunged at her she decided that discretion was the better part of valour and nimbly jumped to the safety of the car roof where she found her feet made a lovely noise. She ably executed a few steps that Fred Astaire would have been proud of. Ted was enjoying it all immensely, after all, it wasn't his car, but he went out to help. Topper ended her performance with a neat routine down over the boot and on to the ground level where I met her with a few Polos and lured her into her shed. Steve refrained from comment and silently polished her footmarks off with a duster. When he came indoors he made a snide remark to the effect that some older women seemed to need strange pets in their dotage.

Chapter 3

Progress

I desperately started to read goat books but each bit of information I gleaned was disputed by Topper. For instance we went to the trouble of buying a bag of special feed for her with added minerals and vitamins because she was soon taking less milk. It looked like the most expensive muesli. I presented it to Topper who sniffed gingerly at it and backed off.

'You don't expect me to EAT it, do you?' her expression said.

'Come on, Topper, we've got a lot of it,' I pleaded. But it was no good, she preferred cream crackers, cake, biscuits, apples, Polos – oh yes and cigarettes. Lighted or not she was not above pinching one from your fingers and she ate it, filter tip and all.

The books said that goats liked being groomed. I bought a suitable brush and prepared to give her a treat. At the first stroke she leapt away, bleated a protest and trotted off to a safe distance. I gently tried again, showing her the brush and allowing her to nibble it but no way was she going to be groomed.

Goats like company, I read, and are better kept in twos or more. I at once felt guilty but it so happened that I was asked to look after another goat while its owner was on holiday. The boarder arrived, another female who was called Flossie. If Topper seemed happier with a companion perhaps we might consider getting another one.

I let Flossie into the yard and waited for Topper to show her friendship. Flossie was eager to make friends with Topper and I imagined a happy fortnight for them both. Not so, Topper butted Flossie in the ribs during the first moment of meeting. Eventually, after a day or two of distrust the two of them were to be seen grazing on the same patch of grass in the orchard. Topper saw to it that Flossie knew her place which wasn't anywhere near me, Ted or the kitchen door. I spent the fortnight surreptitiously giving Flossie her milking time and accompanying feed without Topper seeing. Topper spent the fortnight surreptitiously creeping round after me to make sure I gave Flossie nothing at all. I developed a huge guilt complex and started nervously jumping every time Ted came near me unexpectedly.

When Flossie went home Topper showed no signs that she missed her new friend. It appeared that Ted and I were all the company that Topper needed. It was around now that Ted started laying the foundations for the first pig house to the sound of the concrete mixer so neither of us went out much, and certainly not together. Topper had our company all day.

One evening Ted read in the paper that some prefabricated houses were being taken down and were offered for sale.

'I'll certainly ring up about them,' Ted sounded enthusiastic.

'Mmmm,' I murmured as I struggled with a knitting pattern, either the designer was at fault or my tape measure had stretched and things were not matching up.

'It's just what I want for the pig houses.'

'Prefabricated houses, the sort people live in?'

'Lived in,' corrected Ted. 'They're in sections you see, I can buy them in bits and put them up as I want them.'

'How many sections would you need?' I put down my knitting, I didn't like the way the sleeve was longer than the front, well, not that much longer.

'I'll need two or three houses.'

'Whole houses? Ted, prefabs have rooms and windows and things – we only need walls. How on earth can you put a house together?'

'They're sectional, I can use the sections however I like. I'll need extra supports of course and I'll need to revise my plans, but prefabs had insulated walls which is just what is needed for the pigs.'

From then on Ted spent his evenings drawing out floor plans and then altering them and showing me how the dung channels would be and how the feed passages would have gates across them until my head was swimming. I found it didn't help the progress of my knitting which by this time looked as if it would fit a one-armed chimpanzee. Ted ordered the prefabs to be sent to us and one day in early spring a heavily loaded articulated lorry drew up alongside our field completely blocking the narrow lane; it practically blocked out the sky.

Ted had recruited help from the Fox and Goose the night before and a motley band of curious volunteers stood open-mouthed the other side of the fence. As soon as the engine stopped Topper sauntered up to join us. She was happy to see the men because visitors generally meant tit-bits for her but she was eaten up with curiosity as well. The helpers eventually found tongue.

'Gawd help us, Ted, you'm got some stuff here, boy, where be us going to put it all?'

'We'll pile the sections on top of each other just here,' Ted directed. 'If we put the same sized pieces on top of each other it'll make things easier.'

'I'm glad it'll make things easier,' said one, 'Cos none of it looks easy to me. I reckon us is surely going to need that there refreshment afore us is done.'

That there refreshment was a couple of crates of beer which Ted had ostentatiously carried out of the Fox and

Goose the previous evening as bait for would-be helpers. It had done the trick, there were seven men of various shapes and sizes standing staring at the load. I certainly couldn't believe the enormity of the job of unloading, let alone the rebuilding of it all into pig houses.

Topper settled down on a tump of grass from where she had an uninterrupted view. The lorry driver unroped his load with great aplomb, much as a magician would reveal a rabbit. The unloading commenced, the men struggled and heaved coping with heavy sections, doors and windows. Ted tried to work and supervise at the same time.

'Put this bit on your pile, Jan, we'll keep all the windows together on your heap, Bill, and the three-foot bits over there and the four-foot sections here.'

I was reminded of the part in *Wind in the Willows* where Rat was sorting out weapons for the raid on Toad Hall:

'Here's a gun for Badger – here's a gun for Mole – here's a gun for Toad. Here's a cudgel for Badger . . .'

It all seemed to go on for ages, the beer in the crates was drunk and glasses were strewn about in the grass. Suddenly there was a shout.

'Dang me if that there goat of yourn isn't a-drinking my beer!' True enough, Topper had tried the dregs and found them to her liking. Her head was pushed in a glass as far as it would go and her pink tongue was working overtime.

'OI, Topper!' I called. 'What d'you think you are doing?' In answer she trotted over and leaned her head up against me which meant she wanted me to rub between her horns. I obliged and as I did so she hiccuped. Her legs folded beneath her and she slumped to the ground where she remained for the rest of the afternoon deep in slumber.

During the operation a council van came along the lane and, of course, was unable to pass. No way could the lorry move to make room so the driver was a captive audience.

'What you party up to?' he enquired pleasantly.

'It's like this, Charlie, the chap who have bought this

place is going to build holiday cottages here with these bits and pieces.'

Charlies eyes widened.

'E never is, is 'e?'

'Thass right enough, you'll have to book early an' all, lad, if you want a holiday. We're all getting a free weekend in return for helping.'

Visions of hysterical planning officers flitted across my mind but I enjoyed the fun. At long last the empty lorry drove away and aching backs were painfully straightened up. We all withdrew to the kitchen for a cup of tea and the last of the beer. I tuned my ear into the Cornish dialect.

'You party in the corner want the sugar?'

'Tellee what, boy, you can pass I another bit of that kek, thass some good stuff, Missus.'

I was glad that my cake was approved.

'I bet ole Charlie is a-rubbing his head and wondering about the holiday village you're building here,' grinned Jan.

'He'd know you were joking, wouldn't he?' asked Ted.

'Dunno as he would, he's a bit mazed some of the time but other times he can catch you out as you pass by. Do 'ee remember his ole dad, Bill?'

'Ah, I do that, he used ter hand out thick ears to us kids freer than chapel hymn books.'

'An' he was a mean chap, too, my dad used to say he was tighter than a bull's ass in fly time.' Jan was enjoying reminiscing and Bill was joining in.

'Mind you, Jan, he was a worker even though he was thinner than a rasher of wind. To see him going up a ladder at hay time you'd have said his backside was like a pair of kippers clapped together.

'Oh well, best be making a move back to her indoors or there'll be trouble. Well, Ted, all the best to you and I hope you get on with your bits and bobs you got out there, if we can help you at all you've only got to say. Best keep him well fed, Missus, he's going to need it.'

When everyone had gone and Ted had sunk into a hot bath, I went out to collect any glasses or bottles left in the field. Stark piles of prefab sections leered at me in the twilight. One heap had an irregular lump on top: it was Topper sound asleep, and I was certain I heard her snore.

Chapter 4

Reinforcements

It was a mistake to have livestock before getting the pig houses up. Ted and I realized it but neither of us voiced our thoughts until the first of the concrete bases was laid. This marked progress we felt, now anyone with half an eye could see what we were doing. They could also see what Topper had been doing: there for posterity was a trail of trotter marks diagonally across the otherwise smooth surface, and that was when Ted put into a few well-chosen phrases what he really felt about free range lifestock. I can't recall his exact words but I know Topper's parentage came into it, as did promises for her immediate future which didn't seem to allow time for her to become a milking goat. I seem to remember that John and Sally came into it as well.

In fairness Ted had put up with a lot. When he'd first marked out the ground for the concrete base he'd used marker sticks and string. Topper either rubbed her head on the sticks or walked straight through the string taking it with her. Then she discovered that when bags of cement are really fresh, they are warm to the touch so she loved to sit on them. This wouldn't have mattered but her sharp little hooves made holes in the paper bag so when Ted went to lift it cement scattered in all directions.

In readiness for collecting various building materials from far and near we invested in an elderly Landrover. It was slow, and it ran on diesel, at least that's what we were told. Our Landrover didn't run anywhere, she sort of lumbered and we arrived everywhere in a billow of blue smoke. Sally would see the smoke cloud on the moorland road minutes before we arrived and she would have the coffee ready by the time we alighted. By the time we'd parked up at the Fox and Goose the landlord, Sid, had our drinks on the bar.

Ted had to spend quite a lot of time underneath the Landrover doing mysterious things to her interior. The sight of Ted's feet sticking out from under her intrigued Topper no end. She would kneel on her forelegs and wriggle as far underneath as she could to join him. She huffled over his face first and would then nibble at his hair or do butterfly kisses over him and stare into his eyes. This ultra-feminine treatment would prove too much and Ted would jerk away, banging some part of his anatomy in the process. He would then shout and curse which had the effect of making Topper retreat hastily. Unfortunately, the backward-curving horns that she now sported would hitch in Ted's overalls and there'd be a ripping noise and renewed curses.

It wasn't long before Ted realized that he was going to need help. I'd realized this ages before and wondered how on earth we would manage because we were in no financial state to pay for labour.

'I shall need a bit of help,' Ted stated one evening as if this was an original thought of his own. 'I was thinking, we could ask Cliff down for a bit of a holiday.'

'A holiday?' I queried. 'Hardly that.'

'Well, he's said that he'd love to see the place and there he is stuck in Bristol with nothing to do, I'd be doing him a favour really.'

'Is Cliff up to helping you? I thought you said he gets asthma a lot.'

'He does, but in between he's fine and the fresh air here would do him a power of good, and a few square meals.' Ted went to the phone and started to dial. 'I don't suppose Cliff has ever been in the country, it'll be an experience for him.' There was a broad grin on his face as he waited for Cliff to answer.

I had met Cliff briefly, once, and I felt he was a city man used to the shops and pub nearby, not to mention street lights at night and buses at hand. Just how he would get on with us in the middle of Bodmin Moor I didn't dare to imagine. Still, Ted had known Cliff for years so presumably he knew what he was doing. Certainly Cliff would enjoy seeing Ted again and I could make sure he had regular meals which would be good because Cliff at this time was out of work, due to his poor health, and was in a bedsitter due to the break-up of his marriage. I also knew that Cliff was quite a character in his own way with a heart of gold, no money sense and a strong streak of Andy Capp about him. I awaited his arrival with interest.

Ted met him off the coach in town and brought him back in the late evening when it was dark, so he couldn't see the piles of building in the field. Cliff stood in the doorway, a slight figure with over-long hair that waved prettily round his ears. I felt he only needed a violin to complete his appearance.

'Hello my love,' he greeted me. 'Where the hell am I? You can't see nothing around here, no lights, no houses, no nothing. How d'you know when you're home?'

I ushered him in and fed him, protesting.

'Honestly, love, I can't eat all this. I don't eat much at home but a beer would go down fine,' he said appealingly to Ted who produced one for him. Cliff admired our cottage, he liked the beams and the open fireplace, he liked the cosiness, he said. He'd never been to Cornwall before and was looking forward to seeing it in the daylight. He and Ted had a long talk about our plans for the pigs.

'You said something about prefabs on the phone,' Cliff remarked.

'That's right,' Ted nodded. 'You see, I was able to get

some quite cheaply and I'll use the sections.'

'Good idea, how many were you able to get?'

'Six.'

'Six? Good God, they'll take some putting up won't they? You getting a builder in or casual labour?'

I decided to go into the kitchen at that point. When I returned Cliff was gazing into the fire with a bemused look on his face.

'Come down for a holiday, he said, the break'll do you good, he said, the country air will set you up, he said. I might have bloody known.'

'Come on, Cliff, you'll enjoy it – honestly – we don't have to lift anything. We'll load the sections on the Landrover, drive them across the field then the tractor hydraulics will do the rest for us. We'll just bolt things into place.'

'Oh yeah, I was always good at Meccano. I'm really going to enjoy myself, I can feel it in my water.' Cliff lit a cigarette with fumbling fingers; as the smoke rose past his eyes he wheezed gently. The wheeze gradually grew into a full-scale cough and his face turned red, to my fascination. Cliff coped with all that, keeping his cigarette in his mouth the whole time. Ted eventually calmed him down and we all went off to bed.

Next morning Ted and I were in the kitchen which is beneath the spare bedroom. We heard a bump and some coughing.

'That's Cliff starting up,' Ted told me. 'He sits on the edge of the bed and has his first cigarette of the day.'

'Should he smoke if he gets asthma?' I asked.

'Course he shouldn't, but he does.'

There was the sound of feet moving across the room and another couple of bumps.

'That's him putting on his trousers, I reckon he'll open the curtains next.'

'He'll see the piles of prefab and rush off home,' I gloomily forecast.

We heard the curtains being drawn followed by a silence, during which Ted and I stood looking at the ceiling.

'What the –' cough, cough, '– bloody hell is all that?' cough, cough. The coughing came nearer and nearer with the sound of feet pounding over stairs. A wild figure stood in the doorway, hair awry, red face, unfastened trousers held together with one hand and a cigarette in the other.

'Here, Ted, what the blazes is all that out there?' Cliff spluttered. 'All those bits are enormous, hundreds of them. How on earth can just two of us . . . I mean to say . . . you said to give you a hand, not build the wall of China all over again.'

'Look, have a cup of tea and calm down. If you really don't want to stay you don't have to. Of course I'll take you home if you don't like it here.' This was Ted being placating. I poured out a strong cuppa and Cliff poured most of the contents of the sugar bowl into it, saying that sweet tea was good for shock. He then proceeded to place his cigarette absent-mindedly on the plate where Ted's toast was. He tried to tuck his shirt in, having done the buttons up all wrong, but appeared not to notice the uneven fronts. He retrieved his now buttery cigarette and drew on it, ignoring the sizzling noise that resulted.

'It isn't that I don't like it here,' Cliff moaned. 'It's just that I feel inadequate. I'm not used to the country even, so much noise this morning and I thought the country was quiet and peaceful.'

'Noise?' I enquired, raising my eyebrows at Ted who shook his head.

'Yes, noise. Cows I think it was or bulls, across the road.'

'Oh them,' I exclaimed, enlightened. 'They belong to our neighbours.'

'Why don't they keep them at their place then?' demanded Cliff. 'And what about the banging noise?'

'The – er – banging noise? Oh, I know what you mean. That was only Topper.'

'Topper? Who's Topper?'

'Our goat,' I told him. 'She comes to the back door and bangs her head on it until I give her some breakfast.'

'A goat?' spluttered Cliff. 'Here? You've got a goat? Oh my God, why have you got a goat?'

'For milk. At least, when she's older she'll give us milk.'

'Oh no, not goat's milk, please get me out of here before then.' Cliff covered his face with his hands and peered through his fingers at me. 'Does the goat butt people?'

'No, of course she doesn't. She'll love you and you'll love her,' I said reassuringly.

'No, I bloody well won't,' muttered Cliff.

However, Cliff stayed. At first I think he stayed because he couldn't believe what was going on and then he got involved and couldn't bear to miss any of it. Cliff had a strong streak of curiosity, besides which he had a super sense of humour and the ridiculousness of the situation in which he found himself appealed to him no end.

As it happened the weather was awful for a couple of weeks. It would pour with rain for days on end and then, just as the ground started to dry out, down would come more rain. On one of the drying out days Ted suggested, almost as a favour, because Cliff had been a garage mechanic before his asthma intervened, that he might like to have a look at our Landrover with a view to cutting down on the smoke screen. It was underneath that Cliff became more closely acquainted with Topper, much more closely. Force of circumstances, Cliff's soft heart and Topper's gentle persuasion formed a love-hate relationship between them. I could overhear their friendship growing.

'Topper, you daft bloody animal, how can I do this if you're poking your head in my face? Mind what you're doing with your horns or you'll – ow, you bugger, get off. Now look, there isn't room for both of us under here, you'll have to go. Well, if you stay still perhaps I can manage. (Pause.) Who's a lovely girl then, does she love her Uncle Cliff?'

When at last building operations were able to start, Ted and Cliff seemed to spend much time arguing happily about the best way to go about things. I felt they both enjoyed this and there were side bets on who would prove right. I was occasionally silly enough to venture an opinion and this would draw them both together, united in scoffing at my ideas. Topper deserted me for the thrill of the building site. The minute a section was erected she would sit herself to the leeward side and watch closely the antics of Ted and Cliff. She would eat any biscuits I brought out with their many cups of tea and coffee and scatter her own personal currants over any coat left about. Often she would lie on a discarded jacket, covering it with white hairs because she was moulting. When the workers broke off for mealtimes Topper would fit in some grazing, ready to join them again afterwards. The first pig house took shape; it had walls with windows in and the door even sported a knocker and letter-box, what's more it was number 63! A spell of super sunshine enabled the men to start felting the roof. This annoyed Topper because she couldn't get up there with them. She sulked and came to me for sympathy and biscuits.

The haymaking season began. Ted wanted to make one field of hay for our cattle and our neighbours, Mr and Mrs Pascoe Senior and Joe and Meg Pascoe Junior, were overwhelming in their offers to help.

'We'll cut that for you, no trouble at all, little bit like that'll take no time. Don't you think about getting someone in to do it, we'd be ashamed if neighbours couldn't help out.'

Partly to have a break from the strenuous building programme (which despite Ted's reassurances was strenuous in the extreme) and partly to pay the Pascoes 'in kind', Ted, Cliff and I started haymaking. Ted knew all about it, of course, and I'd helped at harvests before but Cliff was like a lamb to the slaughter. Under the impression that haymaking was rather a lark he set off gaily for the field where it was all happening. An hour later, sweating profusely and red from the sun and his exertions, he gasped, 'You know, Liz, I always thought hay was light and fluffy stuff that you chucked about in armfuls. I never realized it came in these heavy lumps.'

These 'heavy lumps' he referred to were the hay bales that he and I had been stacking up in eights, ready for Joe to pick up and get on to the trailer. Ted, being a seasoned harvester, had gained himself a mechanized job with a tractor. He was off on the next field turning hay to dry it in the sun. This did not escape Cliff's notice.

'I see Farmer Ted's got himself an easy number while the likes of you and me slave away over these lumps.'

'They're bales, Cliff,' I insisted. 'And Joe asked Ted to do that.'

'Well, I reckon I'm working harder than him anyway.' Cliff sounded quite smug and pleased. He'd been so much better since he came to stay. For one thing he was getting regular meals, and for another the work programme didn't allow for as many visits to the pub as he was used to at home and as we kept pointing out, the fresh air was doing him the power of good. Cliff contested some of this. The fresh air, he pointed out, was nearly blowing him off his feet. His fair complexion was utterly ruined and he'd grown webbed feet in the first fortnight. Now, during haymaking weather he was sweating himself dehydrated. Despite all this, Ted said he was looking better than he'd ever seen him look.

The second day of harvest we'd just finished breakfast when Ted called, 'Hey, Cliff, Liz, come quick. I can see some bullocks in the bale field.'

I knew what he meant. The Pascoes' bullocks must have somehow got into the field where Cliff and I had spent so long yesterday stacking up bales and the bullocks, having a playful nature, would be knocking them down for the fun of it. I galloped out of the house followed by a puzzled Cliff. I was bound to laugh because he obediently ran alongside me as I headed for the field gate, only after a few yards did he ask plaintively, 'Liz, where are we going?'

Once at the gate we could see Ted had got to the far side of the field already and had at least caught the bullocks attention so that they had stopped playing with the bales. Even so, several stacks were scattered on the ground. Cliff was indignant.

'What do they do that for?' he asked.

'They rub their heads on them and then find they can knock them over,' I replied.

By this time, the three of us were behind the bullocks. We rounded them up and eventually got them out and back to where they should have been.

'Whew,' gasped Cliff. 'Didn't think I could run like that, come to think of it, I didn't know cows could run like that.'

'You did very well,' conceded Ted.'

'I did didn't I?' smirked Cliff. 'Shouldn't have been so brave if it had been boy-cows.'

'That bigger one you cornered over the far side and chased to the gate *was* the bull.'

'Oh yeah, now pull the other one. I heard you say bullocks and I know that's young ones.'

'Well, it's a young bull.' Ted was relishing this.

'Get on,' scoffed Cliff.

'Look at it then. See for yourself,' commanded Ted.

Cliff stared at the animal in question, bending slightly to see better the tell-tale area.

'Oh my Gawd,' he muttered, going pale under his tan. 'I didn't think bulls were allowed out loose like that. They're dangerous animals aren't they?'

'Only if you chase them,' Ted stated, rather unkindly, I thought.

Naturally enough, Cliff encountered bullocks several times and became quite efficient at ushering them along lanes, having made certain that the bull wasn't with them. He remained uncertain about identifying young heifers and bullocks but was on firmer ground when he could see milk bags 'swinging in the wind' as he put it. To be on the safe side Cliff referred to all bovines as bull-cows, but despite his care he had an encounter with a bull again. This time Cliff was on a tractor and had every right to presume that he had the superior position. The bull had taken up his stance at the entrance to the barn where Cliff was to drive in with the trailer to be unloaded. The bull was happily chewing cud in the sunshine with no immediate plans to move anywhere. He was a large animal and effectively blocked Cliff's way. Cliff edged the tractor nearer and nearer until the radiator and the bull's muzzle were almost touching. Ferdinand eyed the tractor calmly and stayed put. Cliff tried shouting, to no avail.

'Get off the tractor and drive him out,' shouted Ted who was waiting impatiently on top of the haystack to pack away the next load.

'Who me?' answered Cliff with one eye firmly on the bull's horns.

'Get on, Cliff, there'll be another trailer along in a minute. He won't hurt you, he's friendly. Get down and wave your arms about a bit.'

'No damn fear,' answered Cliff with feeling. 'If he isn't afraid of the tractor he ain't going to move for me.'

Ted had to climb off the stack to shoo the bull away in the end. He just shouted and waved a bit and Ferdinand moved slowly off, but not before he rubbed his head on the tractor. According to Ted, Cliff sat gripping the tractor steering-wheel tightly with his eyes shut while it all happened. Ted had to shout an all-clear at him to get him going again.

Cliff found that lifting the bales by their strings made his hands sore. To the delight of Joe Pascoe, Cliff turned up with a pair of gardening gloves one evening. Cliff didn't mind how much he got ribbed as long as there was a laugh in it somewhere. Even Cliff was helpless with mirth when he sent a bale up the elevator with one glove tucked neatly under the string. As everyone said, it was a good job Cliff didn't leave his hand in it.

Soon after the hay was safely gathered in Cliff had to return to Bristol for his medical check-up. Ted graciously gave him leave to enjoy a few weeks back in civilization.

'Do you think he'll want to come again?' I wondered.

'I think so.' Ted sounded more confident than I felt. Still, time would tell. Meanwhile, things seemed quiet and even a bit dull. This didn't last.

Chapter 5

Topper Keeps Us Amused

Topper's main objective in life was to gain entry to the house and it was inevitable that she succeeded fairly often. At first we were inclined to think it rather cute as she tippy-tapped on her little hooves across the kitchen, poking her inquisitive nose into everything. Subsequent visits were amusing if somewhat irritating when we found she'd digested a goodly portion of the day's paper or when she'd knocked over the clothes-horse full of clean ironed clothes and then chosen to sit on them. Another 'Topper Incident' took place when my mother had come to stay. Mother is not a countrywoman, she lives in a town flat and enjoys having the marks of civilization around her like pavements, street lights, shops and a pet-free environment. She dresses smartly on all occasions and my normal attire of jeans and wellies pains her somewhat. Her bedroom is full of carefully folded clothes, hairsprays, perfumes and knick-knacks necessary to the smart woman. My heart sank when I heard her call out plaintively, 'Liz, the goat's on my bed. What shall I do?'

'I'm coming,' I shouted and galloped up the stairs. The front door must have been left open, but why, oh why couldn't this have happend when Cliff was in the spare room? I encountered Mother in a stance of helplessness with her hands clasped as if in prayer which, come to think of it, she probably was, seeing that Topper was lying on her mink coat on the bed.

'Topper,' I commanded in my best Woodhouse manner, 'get off!'

Topper regarded us both with the air of someone whose private suite at the Hilton had been invaded. She burped quietly and commenced to chew cud. She did look rather attractive sitting on the pale green bedspread and the dark glossy fur.

'Topper!' I repeated sternly but with less conviction because she looked so settled. She regarded me with interest to see what would happen next.

'What are you going to do?' asked Mother somewhat testily.

I moved round the bed so that I was behind Topper, having the vague idea that if I shooed her off from there she'd also be facing the door. I shooed. Topper glanced casually over her shoulder but didn't move. In desperation I pushed her. She resisted at first but when she realized she was sliding to the edge of the bed she jumped, as only goats can, over the foot of the bed and came face to face with the dressing-table mirror. Horrified, I watched her stretch out her neck toward Mother's expensive array of bottles and jars. I was convinced this was going to be a disaster when Topper recognized her reflection in the mirror as a friend. She huffled on the glass and watched her new acquaintance disappear in a cloud of mist. This sudden quirk in the weather conditions kept Topper's interest for a moment during which I held my breath. Any clumsy move now could be expensive. All at once she lost interest in this elusive friend and stepped toward the bedroom door in a dignified manner. She managed to make it clear that she was going only because she wanted to go. I followed her out and witnessed the comic view of the rear of a goat going downstairs. I breathed a sigh of relief as I shut the front door behind her.

'Is your coat all right, Mum?' I called out.

'Everything is fine,' was the reply.

'Would you like a cup of tea after all that?' I asked.

'No thank you, dear, but I would like to try that bottle of gin I bought this morning. I'll come down and pour it out, if you don't mind. You don't always put enough gin in mine.'

Topper continued to amuse us, if that's the right word, and Ted and I gleefully recounted every episode to John and Sally, trying to sound hurt and injured in order to jar their conscience. We usually ended up laughing together, but there came a disastrous incident which prompted John to buy us two drinks in quick succession at the Fox and Goose.

I was making some pastry and had got to the stage when I'd rolled the first piece out when Ted came in with an 'elevenses' look on his face, so we both stopped for a coffee. Ted was leaning on the working top and we were both gazing out of the window at Topper who was grazing on what would have been a lawn if we'd have time to level it and seed it. As it was, it was only a rough patch with dandelions, dock and goodness knows what else which Topper regularly sorted out. She saw us through the glass and bestowed her amber stare while her lower jaw rotated non-stop.

'Won't be long before she goes to the billy,' remarked Ted.

'Surely she's too young,' I protested.

'No, she's about right, remember we've had her a fair while and she was already a few months old when she came to us.'

'Shall we know when to take her? I mean, I know what the books say but shall we be able to tell?'

'Of course, as a matter of fact I think she was in season last week, you know, the day she was bleating a lot.'

'I thought she was doing that because I went out.'

'She was wagging her tail as well.'

'It doesn't last long does it, I'd better get in touch with the lady who has the goat stud and make sure we can go when necessary.'

'It would be a good idea, I've no intention of wandering over the moor with her on the off-chance we might find a billy.'

'I hope it'll be a nice billy, I shouldn't like her to have one she doesn't fancy.'

'When the time comes I'm willing to bet that any male goat will do,' grinned Ted.

The object of our discussion was staring extremely intently at us as though she could hear what we were saying.

'Do you want a handsome one or would you rather have a good character?' I said to her. 'The two things don't usually come together.'

'You were one of the lucky ones then,' Ted was saying when I saw Topper bouncing toward us.

'Ted, she's going to—'

At that moment Topper jumped at and through the window. I saw her hard head make contact, the glass splintered into huge zig-zag cracks that shot across the whole pane and Topper landed with a crash and a shower of glass on the working surface where her hooves slithered and slipped. She slid along where I'd been making my pastry, quite out of control. As she skidded along she collected the rolled-out pastry, my roller, the flour bag, the weighing scales and a bowl of eggs. She couldn't regain her feet on the slippery surface and only thudded to a halt when she reached the end of the working top and slumped against the electric mixer. Everything teetered on the edge and I watched in horror as the eggs rolled off and plopped to the floor. I couldn't move for what seemed like ages, then Ted and I lurched forward, I scooped up Topper into my arms and Ted saved the mixer. Nearly everything else found its own way to the floor.

Topper gave a soft bleat, she was dazed, how was she to know that there was glass in the way or that it would matter? Now she'd had a bump on the head and a terribly undignified landing and she expected sympathy.

'Poor Topper,' I murmured. 'Are you hurt?'

'Get that bloody goat out of here before she does any more damage,' roared Ted.

I couldn't for the life of me see how she could possibly do any more, but recognizing anger when I heard it I managed to open the door and stagger out. Tenderly I placed her on the ground and looked her over for cuts; she seemed completely uninjured.

'Maa-aaa-aaa,' she said accusingly to me.

'It wasn't my fault, you silly goat,' I grumbled. 'Are you all right?' In answer she trotted off and shook herself, paused to scratch her back with her horns and set about a clump of grass she'd missed earlier. I sighed deeply and returned to the disaster area. Ted was still clutching the electric mixer.

'She's all right,' I reported. 'Doesn't seem to have hurt herself at all.'

'Oh good, I am glad.' Ted was heavily sarcastic. 'As long as Topper's OK perhaps you can tell me what to do with this mixer?'

'Put it down somwhere,' I said. 'And don't get cross with me, it wasn't my fault.'

'Just where d'you suggest?'

The scene was a shambles. Flour had powdered everything, my pastry was in a lump embedded with broken eggs, the squashed flour bag and a box of matches. On the floor were more broken eggs and flour and our two coffee mugs which we'd dropped. Buttons the black cat, always an opportunist, was licking at the eggs. Jess the collie, who seemed to have an eternal guilt complex, was cringing in the corner presuming that she was to be blamed for the confusion. Worst of all, from the point of view of cleaning it up, glass was in with everything.

'Just put it down anywhere,' I snapped at Ted.

'Like where?'

I had to admit I couldn't see a space. I stood helplessly

wondering how, and where, to start clearing it up.

'I can't stand here with it all day,' grumbled Ted. 'I've got concrete to mix.'

'Take it with you, you can use it if you like.' We glared balefully at each other for a minute and then started to laugh.

'God, what a mess,' said Ted and put the mixer on a chair. 'Just wait till I see John and Sally again. I'll give 'em a dear little goat, no trouble at all, make a lovely pet indeed, I'll tell 'em in no uncertain manner.'

'We'll have to get the window replaced first and foremost, it might rain.'

'I'll see about that, I'll measure the glass and phone up in a minute.'

'OK. I'll make a start here, only I don't know where.'

I spent hours in the kitchen, it seemed, and I was getting up from the floor having washed it over for the third time when I was aware that I was being watched. Topper had her front feet on the kitchen window-sill and was staring in through the frame interestedly.

'Don't you ever think of coming in that way again.' I warned.

'Maaa-aaa-aa,' she answered and proceeded to nibble at a pot plant which had somehow survived her earlier visit.

'No,' I said firmly and removed the plant minus one leaf, but as I did so she helped herself to a Brillo pad, got down from the sill and wandered off with it contentedly.

A few days later I had imparted the gruesome details of Topper's latest escapade to Sally who tried to look terribly sympathetic while giggling her silly head off.

'And that evening,' I continued relentlessly, determined to get sympathy, 'I had to find something for pudding since jam tart was off, and I got what I thought was stewed apple from the freezer only to find when it thawed out that it was tripe and onions.'

Sally pressed her hand to her mouth and failed to

suppress her mirth. It was true that it wasn't the first time I'd had trouble with my catering. If you live away from shops like I do it's easy to forget an essential ingredient and have to manage without. Spaghetti bolognaise with macaroni isn't too bad but macaroni cheese made with rice lacks something, macaroni I think.

'Why don't you get stuff from Ray the baker when he calls?' asked Sally, recovering from her hysteria. 'He sells lovely doughnuts and things which John and I have as a dessert.'

'A baker that calls?' I asked her. 'Who is he, where and when.'

'He comes on Fridays to me. I thought you'd know him.'

'No, I don't but I'd like to. Why on earth didn't you tell me before?'

'I honestly thought you'd know about him.' Sally was apologetic. 'I'll tell him to call this week, shall I?'

'Please do, and wipe that silly grin off your face,' I told her sternly.

'Sorry, Liz, but tripe and onions for pudding, oh dear!' She was off again.

When the baker's van drove into our yard a couple of days later, I was quite excited. As I came out of the house Ray was opening up the rear doors to reveal an array of goodies. So interested was I that until I felt a tug at my jacket I had forgotten that Topper would be just as interested in the visitor as me.

'I didn't know you people kept livestock yet,' said Ray.

'We don't,' I replied, 'Well, just one goat, cats and dogs.'

'It's got horns,' he sounded doubtful. 'It's a billy, is it? Does he butt?'

'She's a she, no she doesn't butt, she's ever so friendly.' To prove my statement Topper chose that moment to jump into the back of the van, landing on packets of biscuits, bags of buns, a tray of crisps and a box of tomatoes. From my screech of alarm Topper realized she'd committed a misdemeanour and uneasily shuffled her feet amongst the vulnerable goods. For a moment I was helpless, to try and shoo her out would only drive her further in. Suddenly she caught sight of a French loaf poking out from the shelf above her head and she started to nibble at it. I reached over her head and grabbed it. Topper's nose followed it and I was able to lure her out of the van and into her shed where I closed the door and left her to finish it. I came back to face the baker.

'I really am most terribly sorry,' I began, then I saw that he was grinning.

'If you could have seen your face when she jumped in,' he laughed.

'I should have expected that, I just didn't think. I'll make sure she's shut in next time you come, that is, if you'll come again?'

'Of course I'll come, worse things happen at sea. Haven't had a goat in the van before though, wait till I tell my other customers that.'

Yes, I thought, word of Topper's escapades travelled fast and I was getting to be known as the lady who has a crazy goat. Ray and I proceeded to sort out the damage. It wasn't as much as I'd feared and I only ended up with a couple of extra packets of biscuits, a few crisps, some squashed tomatoes and a packet of buns, one of which had the clear imprint of a hoof in it. That bun I was keeping for Sally when she called in later for a cup of tea.

Strangely enough, that was the day Topper came into season. I'd let her out of the shed when Ray was safely gone and I noticed she was staying very close to the house. She bleated every time she saw me so I thought that her shoplifting and subsequent feast of bread had given her indigestion. I rubbed her head and gave her some Polos which I thought might help a goat with wind. As I rubbed her head I saw that her fan-like tail was twitching from side to side continuously. When Ted appeared she ran to him bleating pitifully.

'What's all this then?' he asked her.

'Maa-aaa-aaa-aaa,' was her reply.

'Aha, you want something I can't give you,' he grinned. 'Best phone the stud lady and tell her we have an impatient bride-to-be.'

In a flurry I phoned and made sure it was all right to bring Topper that afternoon. I felt like the bride's mother as I rushed to groom Topper who didn't want any of that, thank you very much, and couldn't anyone understand what she really needed? I was just changing into some respectable jeans and a clean shirt, much to Ted's amusement – 'Wearing a hat are you?' – when John and Sally turned up. John had a half-day off work and was thrilled to bits to hear that they were to be in on the nuptials.

'Can we come too?' pleaded Sally?

'You and Liz can sit in the back with Topper, then,' said Ted as he folded the seats down in the back of the car.

'All right, we know our place, don't we Liz? John's best man and you and I'll have to be the bridesmaids.'

'Should we have flowers, d'you think?' I asked Sally.

'Stop your prattle, Missus, and find the bride,' commanded Ted.

'Find her? She was here a moment ago. Where has she gone?' We all looked about us but no goat was to be seen.

'Shy, perhaps, last-minute nerves?' murmured John as we scattered, calling her. Then a workman appeared at our gate. He was from the building site at the Pascoes where a new bungalow was being built. He had a grin from ear to ear.

'Lost your goat?' he said.

'Yes, we have, have you seen her?'

'Seen her? Oh yes, come with me and I'll show you. You can see from this corner.'

We all followed him and he pointed down the road. The youngest builder was backed up against the unfinished wall of the bungalow and Topper had her forefeet planted firmly against his chest. She was nibbling at his beard. It was obvious that he couldn't do anything about it. The third worker was standing by and laughing his head off.

'Your goat arrived about ten minutes back,' explained the man who'd come for us. 'She's taken a shine to young Ginger and won't leave him be. We tried shooing her off but she wouldn't budge.'

'Er, I'd better fetch her,' Ted said pulling a piece of baler twine from his pocket.

'I don't think that'll work,' I said doubtfully. 'She looks pretty settled.'

'She is that,' the workman agreed. 'Seems to be after something. She on heat or in season or whatever goats do?'

'That's right, we were just going to take her to the stud.'

'She must have thought Ginger's beard made him look like a billy,' the man chuckled and continued 'Wait 'til I tell Ginger that, he won't live that down in a hurry. Fancies himself with the ladies does young Ginger.'

In the end, I drove the wedding car to the bride and she was unceremoniously bundled into it. We left poor Ginger much relieved but being teased unmercifully by his workmates.

A short drive, during which Topper bleated out her sorrows for everyone to hear, brought us to the stud. The minute we opened the car doors Topper hopped out eagerly, sniffing the air with an obvious anticipation.

The billy brought forth was smaller than Topper but she hardly bothered to look at him before she swung her hindquarters round to him. He quickly mounted a shamelessly eager Topper and then, or so I thought, he fell off.

'There,' said the goat lady. 'That's once, we'll leave them a minute and let him try again as it's her first time.'

'Is that it?' I was incredulous.

'Goats are very quick,' she informed me.

John and Sally came along the path, they'd hung back not

wishing to seem nosy.

'You've missed it,' Ted told them smugly.

'We've only just got here,' protested John.

'Goats are quick,' Ted grinned.

Topper and the billy were gazing at each other, they touched noses very briefly and the male turned to her tail end. He mounted once more and just as quickly as before he got off.

'Gosh,' I heard Sally whisper.

Topper had no further interest in the billy and she turned her attention to a nearby rose bush. Sally and I tempted her back to the car with Polos. As the three of us sat there waiting for John and Ted to settle up the stud fees, we became aware of an aroma around us. The smell of a billy is distinctive and strong, it's responsible for the general opinion that all goats smell. Topper does not smell normally but after her mating she did.

'I don't think much of your husband's after-shave,' giggled Sally to Topper.

On the way home Topper stuck her head through an open window, well, halfway, and watched the passing motorists. Reactions varied from total disbelief to mad waving and grins. We stopped at the Fox and Goose where Sally and I sat outside with Topper on a lead. Topper enjoyed receiving congratulations and the whole thing started going to her head when Sid the landlord came out with an ashtray of cider for her. She lapped at it eagerly before giving her full attention to some crisps that someone else had offered. She prinked and preened herself as various people, attracted by her presence, came up to speak to her.

'Are you the goat that jumped into the baker's van? I heard all about it. I did laugh mind you, wish I'd seen her. Isn't she lovely?' And then—

'Would this be the goat that jumped through a plate-glass window? My husband told me, he heard about it in the pub. Bit of a character isn't she?'

‘I hope you don’t mind me asking but are you the lady that lives near Dozmary Pool? I often pass that way and I usually have to stop for your goat, she likes it in the road doesn’t she?’

One way and another it was quite a wedding breakfast. At last, Ted, Topper and I arrived home where Topper slumped down on the back door mat and went fast asleep. I felt like doing much the same but I had one more job to see to.

‘It’s the bride’s mother that has all the work,’ I grumbled to Ted as I swept up Topper’s little currants in the car with the dustpan and brush.

‘The bride’s father feels it in his pocket,’ Ted said. ‘I felt I had to buy a pint for the chap who always has to stop for Topper in the road, as well as umpteen drinks for you and Sally.’

‘Umpteen?’ I questioned.

‘And crisps. I thought you didn’t like them.’

‘I don’t, Topper does.’

‘Oh well, in a few months’ time we could have doubled our goat herd.’

‘I bet she’ll have the prettiest kids. Kate’ll be bound to love them, do you think she might have twins or triplets even?’

‘What, Kate or Topper?’

‘Topper, of course. Not Kate, well, not yet. I wish we’d had Topper when she and Steve were learning the facts of life. What a lovely way to find out.’

‘Bit late for them, now. Into drugs and glue-sniffing I suppose. That’s what Sid said this morning, as soon as you get kids off one pot they’re on to another.’

Chapter 6

The Expectant Mum

A goat takes about five months to kid and during this time Topper and I became even closer. It was increasingly difficult to get in and out of doors without her. I reckon she was making up her mind for an indoor confinement. When visitors came who were unused to her ways my conversation ran something like this:

'Yes, that's our goat. Her horns? Oh no, she wouldn't dream of hurting anyone, she only uses them to scratch her back. Yes, she's always been able to open the gate – she pushes at the latch with her head to run between her horns. Your car? It'll be all right there, who told you she jumps on car roofs? John and Sally? Oh did they. Well, she only did that sort of thing when she was little. Now, if you're ready to come in I'll open the door and if you edge past me I'll be able to stop her slipping in. Yes, she loves to come in but she can't if I stop her and if you're quick. She's chewing your skirt? Sorry, it's just that she loves bright colours, I'll wash that bit off for you when we get in. Right, I've got her, you pop in, don't open the door too wide or she'll . . . she has. Never mind, I'll get her out, she won't hurt you but you sit on the deep freeze if you'd rather. There, she's out. You stay there a minute and I'll sweep these up before you step on them, it's funny but she only ever does that indoors if she gets excited. Still, they're easy to sweep. Yes, they're just like currants, aren't they?'

I felt I should make a record, although I'm sure Joyce

Grenfell could have done it better.

When Topper got to the stage in her pregnancy of looking as if she was wearing pannier bags Ted began preparations for the nursery. He started to make a lovely little partitioned bit inside the big barn. His efforts didn't go unnoticed. From inside I could hear,

'For goodness sake get off, Topper, how can I hammer if you're rubbing against my arm? Don't chew that and don't settle down there, I'm going to be sawing it in a minute.' I ventured in.

'How are you getting on?'

'Am I glad you've come, can't you take her for a walk or something, she's being a damn nuisance.'

'She's helping, look, she's picking up your screwdriver for you.'

'Just get her out of here, that's all I ask.'

'Come on, Topper, coming for a walk?'

Often she would come along the road with me and on to the moor, but she wouldn't go there alone. We set off gently as befitted a very pregnant goat. Potter my elderly sealyham decided to come too. At his age a gentle walk was all he'd contemplate. As we ambled toward the moor gate a blur of black and white tore up behind us and Jess the young collie joined us. Being young she is full of energy and a bit bossy so she quickly rounded us up into a tidy flock. Buttons the black cat had been keeping watch on a mousehole at the laneside but Jess decided he'd be better employed coming for a walk with us. Buttons allowed himself to be organized, the mouse would be miles away by now anyway.

The early spring sunshine was warm, larks were singing and a stroll seemed to be the only thing I could possibly think of doing. The dogs were sniffing busily and Topper was delicately picking at titbits in the verges. Any thoughts I may have been thinking were drifting aimlessly around in my head. Then I became aware that I could hear a vehicle in the distance. It was out of sight around the corner behind us but it sounded a heavy thing of some sort. I called the dogs and Jess rushed to me enthusiastically, but Potter is deaf and couldn't hear me or what was coming and he remained standing in the middle of the lane. Topper had heard and was panicking because normally if caught in such a situation she would leap up on the bank but in her present condition, leaping was out of the question, so she decided to get as close to me as she could. Her rush at me coincided with me bending down to grab Potter's collar. I stumbled and sat down in the road beside him. Topper stood right by me with her front feet firmly planted on the bottom of my anorak. Jess put her front paws on my shoulders and licked my face, she thought this was ever such a funny place to sit down.

I couldn't move, I was stuck there when round the corner came, slowly thank goodness, a coach full of holiday-makers. They were idly gazing out of the windows wondering, no doubt, what people did out here on the moors. Now they could see for themselves. Moorlanders sat around in the middle of roads with their animals for company, that's what they did, and they could go home and tell all their friends. The driver braked and grinned broadly. I was aware that people were standing up to see why they'd stopped and faces were pressed against the glass of the windows. I tugged desperately at my anorak and managed to make Topper budge. I got to my feet out of the range of Jess's tongue. The moor gate was only a few yards away; if we ran there we could easily blend into the landscape among the gorse bushes and I could escape all those faces. Unfortunately I hadn't taken into account Topper's love of an audience. As soon as the driver had stopped his engine she had trotted back to the coach, looking eagerly up at the windows. The door slid open and several laughing passengers scrambled out with cameras at the ready.

'Please don't go away,' called one lady. 'I'd love a picture of the goat.'

'And the dogs,' added a little girl.

The dogs deserted me to welcome these friendly people. I tried to wipe my face which was still wet from Jess's licks. I found a paper hanky in my pocket but then discovered it was dirty because I'd used it to wipe the clothes line that morning. Hastily I stuffed it out of sight and tried the other pocket but that only produced a bundle of baler twine. Meanwhile the photographers were trying to group the animals together, they all decided to cluster round my legs.

'That's a lovely picture,' said one, clicking happily.

'Charming,' agreed another.

'Can I give the dogs a biscuit?' asked the little girl.

I smiled weakly. I'd never get away at this rate. The dogs were drawing in their ribs, trying to look starved, and Topper made a long neck to get her share. After dozens of biscuits had changed hands the driver at last said something about having to get on and passengers started to climb back into the coach. I heaved a sigh of relief. Then up over the bank who should flutter into sight but Topper's friend, Peggy the hen, a shiny brown bird who loves biscuits as much as anyone. Peggy roosts in Topper's shed and has developed the habit of riding on Topper's back. Quickly she cleaned up a few crumbs from the road and then flapped onto Topper's back where she cleaned her beak in the fur. There was a flurry of movement in the doorway of the coach and a couple of happy snappers were back in business. At last the exhibition was over. I called to a goat reluctant to see her friends go and I whistled the dogs and positively galloped to the moor gate where the coach passed us, bristling with waving arms. I waved back. Following the coach and until now unnoticed was a small van. In it, laughing delightedly was a nearby farmer who had obviously witnessed the show. He slowed down and called out of his window.

'What's it like to be famous then?'

'That was awful, I'm all hot and bothered,' I told him.

'I expect they'll put you on the coach tour itinerary from

now on, you know, "See the wildlife of Bodmin, Liz Potter and her menagerie", it'll go like a bomb.'

'Thank you very much,' I replied and turned to the blissful emptiness of the moor. No doubt that story would soon spread; I was rapidly gaining the reputation of an eccentric, thanks to Topper.

The last days of Topper's pregnancy were a strain, on me. Topper was fine but I was in a greater state of expectation than she was. I was the one who opened her shed door each morning wondering if I was to see the new family, only to be greeted by Topper with Peggy perched on her back. This friendship was rather charming except that it resulted in Topper's lovely white fur getting somewhat stained down the sides. Topper would sway past me with her passenger aboard and Peggy would flutter off when they arrived at a suitable point.

Ted was working hard as usual, building the pens within the pig house that he and Cliff had erected so we hadn't had much social life, being too busy during the hours of daylight and too tired come the evenings to go anywhere. When Sally rang up to ask us to a meal at their house I twittered a bit.

'I'll have to ask Ted,' I said. 'And Topper is due any time now and—'

'You must come,' Sally insisted. 'You're both working hard and you should have a break. Peter and Mary will be here and they want to meet you both, they used to live near us and I'm sure you'll like them.'

'It would be nice to come, it's just that Topper might kid any day now and I—'

'You jolly well come, I won't take no for an answer, you can't stay and hold her hoof all the time.'

'OK Sally, thanks for asking us. We'll come.'

'That's better. Come about eight and we'll have Chinese. See you.'

When I told Ted he was pleased. I'd been afraid he might chunter about having to go out but he sounded happy at the idea.

'Smashing,' he said. 'Time we had a decent meal.'

'What d'you mean, a decent meal?' I asked indignantly.

'Only teasing. John told me Sally cooks super Chinese food sometimes.'

'That's what we're getting,' I told him. 'And their friends Peter and Mary will be there, too.'

'Sounds a good evening. John said his home-made wines are ready to try.'

We looked forward to going to John and Sally's and I decided to wear a long skirt as they were in high fashion then. I sallied forth to the car in a cloud of perfume.

'Whew!' was Ted's first comment. 'I prefer silage.'

'You're just not used to fashionable women' I replied. 'What d'you think of my outfit?' I should have known better than to ask.

'Oh, are you ready? I thought that was your nightdress.' Before I could think of a suitable reply he disappeared into Topper's shed to see if all was well.

'No sign of anything untoward,' he reported so we set off. Half a mile down the road we came up behind some cows idly picking at the hedgerows and in no great hurry to move for us.

'They belong to Freddie,' said Ted as we came to a halt. 'We'd better get them into the next field. I'll go through them and make sure the gate's open and then you bring them on, OK?'

'OK' I weakly agreed, thinking of my long skirt and sandals.

Ted got out of the car and wove a path through the cows. They were milking animals and used to people close to them, so they didn't move quickly as bullocks would have done. At last Ted shouted to me,

'OK Liz, drive them on, I've got the gate open.'

I yanked up my long skirt with one hand and waved the other arm about in an attempt to get the reluctant ladies moving. They glanced over their shoulders and stared at me then slowly ambled, as only a milking cow can when she's in no hurry, toward their field. Ted was tying the hinge end of the gate to its post with some baler twine as I pushed the other end shut against the last cow's backside.

'Watch yourself!' Ted suddenly shouted. I looked hastily down at my feet thinking I was about to step in something. Instead the last cow was, at that very moment, 'Took short'. The results of her earlier rich eating shot through the bars of the gate, splashing my skirt. I said a few words that I hadn't realized I knew and Ted raised his eyebrows but wisely said nothing. I did what I could with handfuls of grass.

Sally was very understanding when I arrived and helped me to clean up in the kitchen, but I could smell myself during the evening as the room got warmer.

Peter and Mary were a friendly couple and Sally's Chinese food was super. We had a hilarious time because John insisted we all ate with chopsticks. Then John produced the first of his home-made wine.

'Try the parsnip, me dear,' he leered over my shoulder. 'But leave room for the elderberry because that's even better.'

We were all warm, full and mellow and the conversation flowed effortlessly.

'I did so want to meet you both,' said Mary. 'I've heard a lot about you and I think you were so brave to take on the goat, I mean knowing what her mother got up to and everything.'

'Have a crisp,' interrupted John hastily.

'No, thanks. I'd like to hear about Topper's mother,' I said. 'I didn't know you knew Topper's mother, Mary. Do tell me about her.'

'Oh yes, we lived next door to John and Sally. Your goat's mother was Fuschia and she used to eat our roses, couldn't keep her in anywhere, she got through fences for a pastime just when she wanted. Oh yes, we knew Fuschia all right.'

'Are you sure you wouldn't like a nut?' persisted John.

'No, I wouldn't. But I'd love to hear about Fuschia,' said Ted, settling back comfortably.

'Oh Lord,' murmured John and sank into a chair clutching the nuts to him.

'Well, I think the best story was when she went to that wedding,' began Mary.

'True,' agreed Peter. 'Although when she got on the roof and peered in the landing window and terrified your mother was pretty good.'

'I like telling the wedding one best,' decided Mary. Sally came in from the kitchen.

'Whose wedding?' she asked.

'The one Fuschia went to that time, remember?'

'Er, would anyone like coffee?' she said brightly.

'Not till after the story,' I replied relishing the moment. Sally sat on the arm of John's chair and her cheeks grew pink.

'It was like this,' began Mary. 'I was in Sally's garden and it was a hot summer afternoon. We were laying on a blanket half dozing and doing a spot of sunbathing in our bikinis. The church bells were ringing for this wedding and everything was lovely. Fuschia had been trying to lie on the blanket with us but at last she went away and must have decided to see what was going on down the road in the church. She must have noticed people walking down there and followed. Anyway, suddenly round the house came this lady who knew where Fuschia lived.

'Can you come down to the church quickly, Mrs Harris? Your goat won't go away and they're trying to take photographs and Fuschia keeps trying to eat the bride's flowers.'

'Poor old Sally,' continued Mary, giggling. 'You jumped to your feet and you were about to rush to the church just as you were and I said you'd better put something over your bikini first. Well, Sally dashed into the garage because that was quicker than going into the house and she grabbed a terrible-looking jacket, all oily and tattered, and off she goes to collect the goat.'

'It was awful,' said Sally in a quiet voice. 'I go all hot just thinking about it. I had to drag her home and some awful man took photos of me as I pulled her along by her horns.'

There was a silence broken by Mary laughing to herself. Ted caught my eye and winked.

'And knowing all this,' Ted said slowly and deliberately, 'you had the nerve to tell us that goats were no trouble?'

'Didn't they tell you about Fuschia?' asked Mary with wide eyes.

'Well, er, um . . . well,' was all John could manage. Sally was pinkly looking into her glass of wine very intently. I was the one who started to snigger.

'I was pressurized into becoming a goat owner,' I explained. 'And you, Ted, were no better than John and Sally. You'd all fixed it up before I knew anything about it. But the same sort of thing is happening to me now.' Then I told them about the coach incident which reminded me of the imminent birth.

'You don't mean to say you're having more goats?' Peter was incredulous.

'Anytime now,' said Ted. 'Want to order one, make lovely pets?'

'No bloody fear,' Peter said feelingly.

'Bad enough just being next door,' added Mary.

I did enjoy seeing Sally lost for words and Ted enjoyed John's embarrassment. The evening ended in great teasing and laughter and we were quite reluctant to leave, but the thought of Topper was too strong for me.

'We must go, Ted, just in case.'

With promises to let everyone know the news we drove into the darkness.

'Nice evening, wasn't it?' I asked.

'Very nice indeed. John and Sally are so-and-so's, aren't they?'

'I'd have done the same to get a good home for a goat,' I said.

'Oh well, let's find out if we have any more goats.' Ted headed for the shed as soon as he got out of the car.

'Nothing to report,' he said. So we went to bed. Next morning we slept late. I awoke with a headache so I went down to make the tea and get an aspirin. As I waited for the kettle to boil I gazed out of the window at the door of Topper's shed. Topper! How could I have forgotten? I flew across the yard in my dressing-gown and nightdress. I opened the door gently.

'Maaa-aaa-aaaaa.'

Topper was looking a bit affronted, as she does when I haven't a Polo in my hand. Beside her was a miniature replica of herself. Four amber eyes looked at me – no, six. There was another little Topper just behind her. I walked in and Topper bleated again and trotted to me. The two little kids followed, then from the deeper hay in the nursery partition was a rustling noise. A third kid shakily got to it's feet and came toward me.

'Topper, you clever, clever girl. Three babies!'

Topper bleated out her excuses, she didn't know she was going to have all these babies, what was she supposed to do with them all and more important why hadn't I brought her breakfast? Peggy fluttered up on to Topper looking fussy. They all seemed to be looking at me to know what to do next. I rushed out and back into a kitchen full of steam where I'd left the kettle on. I thumped up the stairs and stumbled over my dressing-gown cord, entering the bedroom on my knees with a resounding crash. Ted shot up in bed all bleary-eyed.

'What's the matter? What's happened?'

'Triplets have happened, that's what!'

'Well, I'm blowed!' Ted got out and came to the shed with me in his pyjamas. We spent a good ten minutes going ga-ga over the babies and giving Topper her breakfast and making sure all was well. When we emerged, Ray the baker

was driving into the yard with his van full of goods.
'You two been sleeping out?' he enquired interestedly.
'Topper's just had triplets,' I told him. He thought for a minute, then pushed his cap to the back of his head and had a scratch.
'I best be getting a bigger van, I reckon.'

Chapter 7
Poultry

Peggy is not the only chicken we have but she is the one with an individual public image, the one that gets noticed. She is a beautiful bird with glossy plumage which isn't surprising because she eats Topper's feed and other titbits which she has a habit of acquiring for herself. How we came to have her goes back a bit to one Sunday when we awarded ourselves a whole day off to go and have lunch with my friend Anne who lives in Devon. Anne and I went to school together so we can out-talk Ted and Tom, her husband. Tom is one of those people who can always get hold of things cheap, all sorts of things: paint, Christmas cards, tools, a cottage ripe for renovation or a lawn-mower. This visit was due to the fact that Tom had said he could give Ted various bits of plumbing suitable for the pig houses, so we set off in keen anticipation.

'I only hope that old Tom has the right sort of stuff,' said Ted. 'I don't want gold-plated taps or anything like that.'

'Hardly,' I replied. 'The gold-plated stuff was a one-off offer and he put it all in their bathroom.'

'Mmm, well, you know how Tom gets these things and expects everyone to be in need of it. Look how much paint you ended up with the time he said he'd get you some for the amateur dramatic group to do their scenery with.'

'Don't remind me,' I said. 'That was three years ago and

they've still got enough to paint the QE 2 with.'

However this time all was well and Ted was very pleased with the collection he stowed away in the car. Then we had some home-made beer of Tom's, after which we started to think of going home. I hadn't realized how well the beer was going down until I heard Ted happily agreeing to have half a dozen bantams to take home. Anne, Tom, Ted and I were riotously rounding up the athletic birds and getting them into the large boot of our car when Tom came round the corner of the house.

'Hang on,' he called. 'Here's one more,' and he appeared with a cockerel under his arm. 'You can't leave the old man behind,' he grinned.

'Thanks, Tom,' I said a bit doubtfully. Tom was being very generous with his poultry and I was wondering what the catch was.

'Didn't you want to keep the cockerel?' queried Ted.

'Oh, we've got several, they're a damn nuisance actually, they're fighting and keep flying over the fence and—' he trailed off lamely then gave us a wide smile. 'Well, you've got plenty of room, haven't you?'

'Have we?' I turned to Ted. I think the home-made beer had something to do with his reply.

'Course we have, I'll knock up a house for them in no time.'

On the way home I asked where we were going to put them.

'They can go in the tool shed for tonight and I'll make a house for them tomorrow.'

'What with?'

'There's enough bits and bobs about, no problem.' I was told.

When we cautiously opened the boot as soon as we got home seven pairs of bright eyes looked at us accusingly. We carried them, protesting loudly, to the tool shed where we scattered some food for them and put water in the dog's dish, then we left them to recover from their journey. Anne rang up to know if we'd got home safely and as I spoke to her on the phone which was by the window, I noticed a large brightly coloured bird clumsily flapping at the foot of our hawthorn tree. It was the cockerel. As I watched I could see that he was trying to get up into the lower branches. He failed miserably the first time but on his second attempt he made it onto a branch, frantically waving his wings around in a desperate way to keep his balance. I gave Anne a running commentary. In the hawthorn tree all was not well. I could make out the outlines of the other six birds, they were much further up on the branches and looked quite settled. The poor cock now tried to join his wives. He stretched up his neck and flapped a bit, standing on tip-claw on his lower branch he managed to get his chin over the next one up, but try as he might he couldn't get the rest of him up there. He lost control and plummeted to the ground, crashing through small twigs and making a lot of noise in the process. Anne was speechless with laughter on the end of the phone and Tom came on the line.

'Now look here,' he grumbled, 'I thought you'd be able to look after a few chickens, what d'you think you're playing at letting my poultry roam about at night?'

'Tom,' I bellowed down the phone. 'Did you know they were likely to roost in trees? Just why did you let us have them?' There was a chuckling on the line and Tom gave the phone back to Anne.

'What on earth are you going to do with them?' she asked.

'The hens can stay where they are, they're safe enough and Ted's just picking up the cockerel now, no, it's running away. Ted's close after it, gosh, he's doing a rugby tackle. Ted's got it under his arm and he's coming to the window. I think he's saying something, I'll open the window a minute. Hang on, Anne. Ted? Is the cockerel all right? You did what? Oh dear, never mind I'll wash your trousers – you hurt what? Oh what a shame. Tell Tom what? I can't say that – no – or that either. Sorry Anne, Ted fell down catching the cockerel and got a bit muddy, that's all, and he wanted me to give you a message for Tom only I think it'll be better left till they meet next time.'

The next morning I heard banging and sawing but couldn't see where Ted was working. By the time I sauntered out to look he was coming to fetch me to see his latest creation. There in the corner of the orchard (I don't know why we call it that because there's only a couple of hawthorns there) was a lovely chicken house complete with door and window and snug-looking roof. I stood and gazed at it for a minute.

'Ted,' I said doubtfully, 'are you allowed to do that?'

'Do what?'

'That. You've built the chicken house against the telegraph pole, is it actually nailed to it?'

'Course it is, it won't blow down, firm as a rock that is.'

'It's a jolly nice house for them as long as the post office people don't mind.'

Once the chickens had a henhouse to sleep in they never tried to roost in a tree again and the cockerel was a happy bird once more. During the day they were completely free range so it was anybody's guess as to where their eggs might be. Oh yes, we gave them very good nesting-boxes in the house where they slept and we tried leaving them shut in until they'd laid, but they were missing most of the day in the end so we left them to it. Anyway, looking for eggs was a super excuse for an hour outside whenever I felt like it. In due course we had clutches of eggs hatching out and a proud hen would appear with a collection of fluffy powder-puffs following her. The trouble is, hens can't count and as long as there is one chick to follow them, they never miss the others if they fall by the wayside.

This is where Peggy comes into the story. After a night of heavy rain a hen led into the yard six bedraggled soggy chicks. I ushered them into the barn where they could dry off nicely. Somehow the mother hen found a way out of that unsuitable place for her family but absent-mindedly left one behind. It was the weakest one and when I found it was lying looking dead on the floor. I picked it up, expecting to have to dispose of it, but it's little beak opened slightly as I held it.

I took it into the warm kitchen, it's head wobbling on a thread of a neck. First I tried to dry it but it felt so fragile that I didn't dare rub it. It's minute feathers were stuck to it's skin. How could I dry it without hurting it? I remembered something I'd heard and decided this was the time to try it. Gently I placed the chick in a plastic bag and blew into it so it was like a balloon. I held the top together then I floated the bubble on a bowl of warm water. I was bending over the sink watching my incubator when Ted came in. He looked for a minute and said,

'Sending it into space?'

'It's the only hope for this one I'm afraid,' I told him.

The chick began to get the benefit of the warmth around it. It's beak opened and shut a few times, there was a tiny movement in it's feet and an eye regarded me.

'I think it's going to work, Ted, did you see it open it's eye?'

'Frankly I think you're wasting your time. I suppose I'd better wash in the bathroom.'

'Come on, chick,' I encouraged and poured a little more warm water in the bowl. The tiny thing moved a bit. After a few minutes it seemed to be getting stronger and I realized that I must take it out of its bubble before a claw punctured the plastic bag and we had a drowning on our hands. I wrapped it in a duster and popped it in the warm oven of the Rayburn and kept peeping in to see how it was. By the time I'd made coffee it was standing and leaning on the duster like a drunk. I found a box for it. It seemed no time at all before I was startled by a cheeping noise. Now it was scratching with its feet in the box and looking much better, with its feathers beginning to fluff up.

Feeling like Florence Nightingale I offered it a drop of milk with glucose in it on the end of a paint brush. The drop of liquid was taken and an eager little beak thrust out for more.

Incredibly, by evening, the chick was strong and trotting about in its box quite happily. I put a pinch of crumbs in with it, and an eggcup of water, and left it for the night by the Rayburn. In the morning I could hear its chirps from upstairs.

I hazarded a guess that it was a girl and called her Peggy. She remained indoors for a few more days, progressing further away from the Rayburn. She hopped onto my hand for food and made herself heard above the television. It was after she found her way from the kitchen to the sitting-room that Ted insisted she went back to mum. I tried this but neither of them wanted to know the other. I created a home for her in a cardboard box with a little door so that only she could get in and out, and she remained in the outside room for a bit. Then I put the box in Topper's shed. From that stage Peggy took over. She roosted on Topper's back and took rides around the estate with her friend. Topper accepted her as a companion and they were usually together. Now and then Peggy would join the other chickens for a scratch about but she soon came back to Topper. Peggy also discovered the kitchen window-sill. She was marvellous at cleaning out dishes, especially those containing remnants of overcooked rice pudding. She was extremely good at cake tins and when I dropped a biscuit box on the floor it was Peggy who came in willingly to peck up every single crumb.

As she grew older she did roam a bit further afield, going gleaning with the other hens, but she always spent a good bit of time with Topper and myself.

When we acquired a Muscovy drake from somewhere with a bad foot, he too, chummed up with Topper and

Peggy. His bad foot got better but he decided to stay anyway and soon began to show us he could fly. We called him Boris and as his flying went he certainly wasn't in the jet set. He would scramble up on to banks for take-off and then flap his wings, hoping to gain a bit more altitude from a shed roof from where he got going, briefly, only to land in various comic routines. I think he only did it for fun because he never went anywhere and usually waited for an audience before performing. However his antics impressed Peggy and, having an ambitious nature, she tried to copy him. Until then her flying had been confined to getting on the window-sill but one day she took off from Topper's back and got to the barn roof. Topper had shot off in alarm when her friend had suddenly flapped away like that, but now she regarded Peggy with interest.

'What are you going to do now?' I called out.

'I shall fly down – in a minute,' I imagined her saying.

'Oh yeah?' Topper's expression said.

'I can, you know,' Peggy would have replied. 'Boris showed me how.'

'Come on then,' I encouraged, dying to see the outcome. Peggy sidled to one side, then the other, tilting her head to eye the ground doubtfully. She raised her wings and fluttered, nearly losing balance on the edge of the roof.

'Rrrrrrrp,' she clucked in consternation.

'You can do it,' I told her. 'Come on.' There was more wing movement and Topper backed off a bit.

'Peggy,' I cried in desperation, 'land here,' and I held out my arm. She regarded me seriously for a minute then launched herself into space. Did I imagine her shouting 'Geronimo'? A flurry of feathers landed on my arm and I felt her sharp claws through my anorak sleeve. I steadied her with my other hand.

'Clever girl,' I crooned.

'Taking up falconry are we?' asked Ted from behind me. 'Going to teach her to catch rabbits?'

I ignored the sarcasm, placed Peggy on Topper's back and watched them walk off in stately manner. I headed for the back door.

'Coffee?' I invited. There was a whooshing noise and Peggy landed on my shoulder. Her wing-tip caught my ear painfully and her claws once more dug into me.

'Well, I'll be blowed!' exclaimed Ted. 'Let's hope Topper doesn't learn to fly.'

I mentioned that Kate thought she'd been attacked by the geese. We'd acquired a breeding pair, although that wasn't the term that either Steve or Kate employed to describe them. They turned out to be very good watch-dogs and tended to streak toward anyone they didn't know with necks outstretched, hissing like steam engines. Both youngsters avoided them like the plague and I had even had to escort them to the car once or twice because – 'Mum! the geese are out there!' I couldn't help being amused at the idea of my big grown-up pair being scared of two birds, although they weren't alone as I discovered when I rescued an electricity man. He had come to mend our phone which entailed him going up a ladder to inspect the wires on the pole in the yard. He got up there all right, but by the time he wanted to descend the geese had stationed themselves at the bottom, hissing fiercely. As he started to come down their necks extended up to the third rung. Luckily, I went out at that point to ask if he wanted a cup of tea. I diverted the geese with a bowl of scraps and the poor chap did a hasty retreat.

Steve and Kate, by now, had become resigned to our strange new way of life. They referred to the homestead as 'The Funny Farm' and took delight in repeating the more zany incidents to their friends. Their friends, when they came to visit, were prepared to meet a sort of 'Ma and Pa Kettle' and I was never sure if they were disappointed or fulfilled.

One time Steve arrived unexpectedly with a friend to spend the night. They went upstairs to dump their belongings and came down again looking perplexed.

'Er – there's a box of enormous eggs on the spare-room bed,' said Steve's friend, looking anxious.

'Oh yes,' I reassured him. 'It's all right, it's only the goose eggs, I'll move them.'

'Goose eggs,' he gasped, sounding relieved. 'I've never seen any before. I was afraid they were alligators or something.'

Apparently, later the poor fellow had dropped a shoe under the bed and glimpsed something white and woolly there. He'd gone to Steve with fears of some straying animal only to discover Kate had stored a sheepskin numnah (or under-rug for a saddle) there. When he came down in the morning he was greeted by chirping noises from the warm oven of the Rayburn where I was reviving a clutch of chicks. Still, he was able to dine out on his experience upon his return to civilization.

Chapter 8

Pig Farmers Are We

At last, our building programme was completed, for the time being at least. We were the proud owners of five houses, three we'd put up ourselves with Cliff's help and two we'd had built for us: the purpose-built farrowing house and the weaners' house. The willing Cliff had returned to us on several occasions for his 'Holidays' as he laughingly called them and had become so caught up in the project that he was part of The Funny Farm Madness.

Cliff was with us soon after we'd taken delivery of our first pigs. He'd only been in the house ten minutes before he was demanding to see them.

'Come on,' he urged us. 'Let's see what all the fuss has been about.'

Ted and I proudly escorted him to the converted prefabs he'd helped to erect. The building had its original front door complete with letter-box and street number.

'Oh very nice. Number sixty-three, I see. Do they get much post?' Cliff was still chortling at his joke as we entered the pig house. As he caught sight of the pigs there he gasped.

'Good Gawd, they're big 'uns,' he exclaimed. 'I thought you'd start with little 'uns.'

'These aren't fully grown yet,' I explained. 'They're about three months old and will soon be ready to go to the boar.'

'I've never been close to such big pigs,' complained Cliff, backing away.

'They won't hurt you,' Ted said.

'They can't if I don't get too near,' was Cliff's reply. 'Where do they find a boar then?'

'We've got our own, in the pen next to you.' Cliff moved hastily again.

'Boars are dangerous, aren't they?'

'This one isn't yet,' I said and got in the pen with it. I rubbed the young boar behind his ears and gradually progressed to rubbing his shoulder and ribs. The boar started to sag at the knees and slumped to the floor, exposing his belly to me for more rubbing, his eyes closed in bliss.

'Well I'm buggered,' was Cliff's comment but he refused my invitation to do a bit of rubbing himself. 'I suppose you've given them damn silly names?'

'He's Abraham,' I replied.

'Why?'

'I think it means Father of Nations or something.'

'Very suitable,' Cliff agreed. 'When does he start?'

'He's started,' Ted told him. 'We've some piglets due any day now.'

'Really? Where do they have them?'

'We'll show you the farrowing house next,' said Ted.

'The what house?'

'Farrowing house, you know, the maternity ward.'

'That's better,' said Cliff. 'Use English and I'll understand.'

'The sow farrows when she has young.'

'I'll bet she does,' muttered Cliff.

The farrowing house was a luxury place as far as pigs go. It had fixtures and fittings, lamps for warmth and single accommodation for each mother-to-be. At first I had been worried by the farrowing crates. The word crates suggested a cage or prison but I'd got used to the idea eventually having seen some in use on another farm. I only hoped that our pigs would take to them. When Ted and I had moved in the first sows I practically held my breath waiting to see their reactions. The first one, Princess, ambled in slowly and lifted her snout to snuff the air. She sort of vacuumed the floor and progressed to a gate which was invitingly open to show fresh straw laid ready. Princess walked in and discovered the bars either side of her and made use of them immediately to have a good prolonged scratch. To her delight she found the food trough at the end had nuts in it and she busied herself emptying it. Once finished she sank to her bed of straw and went to sleep.

Would she, I wondered, want to get out? I couldn't bear to think of a pig imprisoned, but to make sure I gave her the option the next morning of being able to escape by leaving the gate open behind her. I didn't tell Ted what I was doing, it was just for my own peace of mind. Princess glanced over her shoulder at the open gate and stayed where she was. I subsequently tried this test on other sows but they all seemed quite contented to be where they were. I had to turn out a sow from her crate once because her drinking nipple had got stuck and her pen was flooded. I had a terrible time getting her out and having done so she created merry hell in the passageway while I cleaned her pen, then she was almost back in before I'd got out of the way so I got my feet trodden on in the rush. To anyone who thinks a crate is cruel I say, come and see before you organize your protest, our sows won't be joining in for sure.

This attitude of mine was reinforced when we had sows coming into the farrowing crates for the second time round. As they entered the farrowing house their pace increased and they would positively trot into the first one with a gate open and slump down giving gusty sighs of contentment. Until they got to these maternity wards of theirs they were living six to a pen in the sow house, having to fight for a favourite patch to sleep on and jostle and shove at feeding times. The farrowing crates provided a sort of piggy Hilton with single beds and your own trough for meals, not to mention the superb facilities for scratching.

The appointed day dawned for our first farrowing and Cliff and I kept popping our heads round the door to try to be the first to see our piglets. In spite of our efforts it was Ted who came to the house and announced the birth. The three of us crept in to see.

Princess's pen had a lamp hanging over it and there was a circle of rosy light to show the first two piglets. They were standing nose to nose wobbling slightly on the daintiest little trotters I've ever seen; they made Topper's look gigantic. Princess was lying on her side oblivious to everything, it seemed, and breathing steadily and heavily. The two piglets made their way to the milk bar and tried to find out what they were supposed to do with the buttons. Princess moved a back leg and propped it comfortably on the lower bar of her crate. Her ropey-looking tail whisked round and round a couple of times and she gave a grunt. Out popped a wet shiny piglet who wriggled and squirmed a bit and made a little squeaky noise. It got on its feet and tried to sort them out. The piglet was still joined to its mother by the umbilical cord, and as it set off in the direction it thought it should go the cord was stretched to its limit like a piece of elastic, and the piglet was halted.

'Shouldn't we help it?' I whispered.

'No, the cord will break by itself soon and the struggling is good for the piglet, helps its breathing and circulation.'

As we watched the piglet straining at its cord, the other end slipped out of the sow and the piglet fell forward on to its snout. It gave a very healthy squeak of disapproval and joined the others at the milk bar. I stayed until Princess had eight little sausages lined up and watched them all learn what to do with the teats. Princess started to give satisfied grunts which made her belly shake and one or two babies fell off, rolling on their sides. On their feet once more, they started sucking again immediately. They soon learnt not to let go so easily and as Princess grunted they went up and down with her belly like a set of puppets. Once they were full they wandered to the warmth the lamp was giving out and collapsed in a heap. A pink pile of pulsating life. We had started pig farming.

We were unlucky to encounter a problem so soon: our fourth sow didn't like her piglets one little bit. She had bitten two of them before we realized just how much she hated them. As she gave birth she heard them squeak and she jumped to her feet and attacked any that went near her head with her mouth open, showing a fierce set of teeth and barking almost like a dog. The piglets were frightened and bewildered so we collected them into a box which we put under a lamp where they, huddled together quietly.

'Damn it all, I nearly ordered some sow tranquillizer from that rep the other day, I wish I had but I honestly didn't think we'd be needing it this soon. An injection now would probably cure her troubles,' moaned Ted.

'It was expensive,' I consoled him.

'It will be, to lose these piglets.'

'Shall I call the vet?' I asked.

'Sunday morning, I'd rather not but I suppose we'd better.'

'We could try beer,' I ventured.

'Try what?'

'You know, I read that bit out to you from a book. In the old days they gave a savage sow beer to get her in a good mood.'

We stood looking at the box of rejections and their furious mother who was glaring at us over her shoulder.

'Well—' Ted hesitated.

'We could pop down to the Fox and Goose and ask Sid for any slops,' I suggested.

'I suppose so.' Ted sounded doubtful.

'If it doesn't work we can still call the vet, it's only one o'clock'.

'All right, we'll try it. Better go right away.'

We arrived at the Fox and Goose with half an hour to spare before closing time. John and Sally were there, surprised to see us.

'Thought you two were on maternity duty this morning?' said John.

'We are,' I whispered. 'That's why we're here, got a bit of a problem.'

'What sort of a problem?' asked John in his usual voice.

'Sssh,' I begged him, mindful of Ted saying he didn't want everyone to know we'd hit a snag so soon. 'We don't want everyone to know.'

'Your secret's safe with me,' John hissed in a dramatic whisper that reached practically all the ears at the bar. He sidled close and put his face near to mine. 'Tell all to your Uncle John,' he invited.

'Do not trust him gentle maiden,' sang someone, and there was general laughter.

'If you and John want to run away together it's all right by me,' Sally said with an artificially bright smile.

'Oh dear, you two at it again?' grimaced Ted. We were getting used to John and Sally having lightning quarrels which as a rule cleared up as fast as they started. Quite often they were very funny because they both went in for the dramatic touch, exaggerating greatly and with lots of gestures.

'Actually I'm a bit too busy this Sunday.' I entered into the spirit of the thing. 'But perhaps next week?' I winked at John.

'Certainly, my dear, but first tell me this secret.' Ted began to explain the pig situation to John and I turned to Sally. 'What's your trouble?'

'John came home very late at night, I'd got worried and when he came in I got a bit cross.' She made it sound so simple but I guessed a grand emotional drama had taken place.

'Well,' I tried to sound cheerful, 'at least he got back OK

even if he was a bit late.'

'I don't know whether to call it late or early,' she continued. 'It was three o'clock in the morning. Do you call it early or late?' she asked me. I didn't really want to answer so I was pleased when John turned to us and said, 'Here, Sal, they're going to get their sow sloshed, shall we go and watch?'

'You're going to what?' Sally's rather grim expression changed to one of incredulity, 'Did you say "sloshed"?'

'We're only here to see if Sid has some slops or any cheap beer and then we're trying it on a savage sow. If you want to come, do.' Ted managed to get a plastic container from Sid with a mixture of slops in it which he tried to smuggle out of the bar, but of course it was noticed.

'You having a bit of an afternoon party?' we were asked.

'You know,' I heard Sally saying to John, 'I think poor old Ted's flipped, it's all the work he's been doing. I ask you, getting a sow sloshed, what next?'

By the time the four of us got into the pig house Sally was reconciled to the idea of what we were trying. The sight of the little abandoned piglets forlornly huddled together in their box, coupled with the harsh barks of the sow as soon as we got near her pen, brought home the serious side of this operation.

'How will you get her to drink the stuff?' asked Sally.

'I think she'll love it,' answered Ted and poured a good measure into her trough. She stopped trying to bite Ted's leg and gingerly sniffed at it. She pushed her snout into it and blew a few bubbles, then she smacked her lips and there was a prolonged noise like children make when they drink with straws, and all the beer was gone. She glared at Ted and tried to bite his trousers so he poured the rest in for her.

'I suppose it was all slops,' muttered John a bit wistfully as we all stood watching the sow for signs that she felt happier. She went on smacking her lips and then rubbed her shoulders on the bars with her eyes half-closed in ecstasy.

Then she looked at Ted again and barked but in a friendly 'Come on old chap, let's have another one' sort of way. Ted sighed and went out of the farrowing house. John, Sally and I looked at each other, wondering what was next. We hadn't long to wait. Ted reappeared with a large bottle of home-made wine we'd had given us some while ago. It only had a little drop left at the bottom.

'I'd forgotten that stuff,' I said. 'Will it be all right?'

'Be all right for a pig,' Ted replied and unscrewed the top. There was a mild explosion and white fumes curled round inside the bottle and started to escape out of the neck.

'Struth,' said John reverently. 'Some stuff that.'

'Rather her than me,' said Sally. 'It won't make her ill will it?'

'I thought it might act as a sleeping draught,' Ted stated and poured a drop into the trough. The sow advanced eagerly and in a minute it was gone so Ted emptied the dregs and we watched them disappear. This time the sow gave a gusty sigh and her legs slowly buckled. She lay on her side looking completely relaxed. Her breathing became noisier until it was almost a snore.

'Now what?' whispered Sally.

'I'll try her a minute,' answered Ted and cautiously began to rub her belly. She shifted slightly to present a larger expanse to him but didn't open an eye.

'I think I'll try a piglet.' Ted picked one out and placed it at the milk bar. It was very enthusiastic, after all it had been in the world for about two hours and hadn't had a drop. Its mother raised her head and sort of squinted down at her teats. I think she thought she merely had a larger teat. She sighed heartily and the baby nearly lost his grip. But he was made of sterner stuff and he took in a good large mouthful and hung on like a little limpet. His mother slept on. Ted added the rest of the litter. All remained peaceful. Ted took a garden chair in with them and read a book until the sow woke up again, to make sure she would still be all right.

Sally, John and I retreated to the house where John went to sleep and Sally and I watched a Sunday afternoon film. It was a weepy and as we were wiping our eyes at the over-happy ending Ted came in to tell us all was well. He looked at our faces and said,

'It's not that bad you know, you didn't need to cry for me, I've had a lovely time reading but it's nice to know you both care so much.'

Sally threw a cushion at him which roused John who came to muttering that he was sorry, really ever so sorry, and then looked so bewildered that we all fell about laughing.

'If you'd said that last night I shouldn't have got so angry,' Sally told him.

'Said what?' asked John, yawning widely.

'Never mind, I'll make some tea,' I said. Just as we were finishing it Cliff came in. He'd been out 'visiting a friend' who just happened to be an attractive young widow with two lovely little girls, and Cliff had escorted them on a picnic 'purely out of the goodness of his heart to carry the hamper'. Of course we all teased him unmercifully before recounting the drama of the sow to him.

'I knew it,' he grumbled. 'I only have to turn my back and everyone goes mad. Giving a pig beer, I never heard of such blasphemy and on a Sunday too.'

Chapter 9
Living Quietly in the Country

The two calves we had from the time of poor Connie continued to thrive. One of them was black and white and I'd named it Whiskey. Topper had considered it her duty to show both calves how to become a successful goat so they were both good at getting under fences and over them. There came a time when our days started with an anxious look out of the bedroom window to see if both or even one calf was in view. Most of the time only one could be sighted, and that was the one I called Sherry who was of a more placid disposition and fairly content to be a cow. Whiskey, on the other hand showed a definite tendency toward escapism, vaulting and the high jump. Ted and I became used to an early morning jog down to the Pascoe's where Whiskey was usually to be found, having a friendly graze with their bullocks. He'd come back as sweet as you please, trotting in front of me to his own field with no trouble, but as soon as I looked the other way he'd high-tail it off again. Ted and I spent ages tracking down how and where he got out, then repairing it only to find him gone again the next minute. After a particularly frustrating day of fetching Whiskey home Ted made an announcement.

'I know we can't afford a lot of new fencing but this is ridiculous. I'll get an electric fencer and stop all this nonsense.'

We hadn't much cash at that time because although our pigs were breeding well we hadn't got any to the stage of pork weight, but we were feeding them of course, and feed bills came through the post only too often. The bank manager was understanding and called me 'Me Dear' but our finances were rock bottom. The only way we could afford to pay for something was to barter a pig. We did barter pig manure to gardeners in return for vegetables and the like. So when Ted made his announcement to spend cash on an electric fencer you could tell the situation was serious. I was terribly relieved because I'd had an embarrassing time due to Whiskey recently. Some friends of ours who lived in Bristol were on holiday in Cornwall and had phoned up to ask if they could drop in to see where we were living and how we were getting on. This 'dropping in' was fine by us as a rule, but these particular friends were in the acquaintance class rather than friends. They were reasonably well off and led well ordered lives, and had thought Ted and I mad to go into farming when we could have bought a bigger house and continued with our jobs and done something sensible with our money like buying a sports car or two and joining a country club.

'We're staying at the hotel,' said Sue when she rang. 'I expect you know it, it's the big one at St Ives?'

'Er, I think I may have seen it,' I murmured. I couldn't bring myself to tell her that since we came to Cornwall we'd hardly had time to go anywhere, let alone the money to stay away.

'We'd love to see you and Ted. It seems ages since you went away and we have missed you both.'

'Do come,' I told her. 'It would be nice to see you, too.' I crossed my fingers as I said it because Ted was making faces at me while I was on the phone. I gave them directions and put down the receiver.

'What d'you ask them for? You know they only want to nose about.'

'I couldn't very well not tell them how to get here, they more or less invited themselves.'

'I s'pose,' grumbled Ted.

When the day came for their visit I did a bit of 'tarting up' in the cottage, a vase of fresh leaves here, the best side of the cushion up there, and I hid away the doggy chair altogether. I banned the cats from indoors and made a fruit flan for dessert on my best plate, warning Ted that a word of surprise from him would earn my deepest displeasure.

'You only use that plate for high days and holidays, why bring it out for them?' he queried.

'Sue always had lovely pottery around and she'll appreciate it. I don't want them thinking we live like pigs.'

'In that case, can you mend these overalls before they come?' asked Ted plaintively. 'I'm positively indecent in them at the moment.' I had just finished sewing them when I heard him shouting at me.

'Whiskey's taken off again, he's making for the Pascoes, can you go and head him off before he gets into their big field? I'd go but I've got the tractor running.' I'll bet you have, I thought as I took off my tidy shoes and donned my wellies. I pulled on my tatty anorak over my tidy 'visitor' clothes and shot out of the back door only to discover that Sue and Dick had already arrived and were just getting out of their car. Their faces had bright smiles on them which faded quickly as I tanked past them.

'Shan't be a minute, got a bullock missing,' I called to them as they watched me disappear. Apparently they remained uncertainly by the car wondering what they should do until Topper came up and worried the life out of Sue by nibbling at her smart coat. Then they heard the sound of hooves on the road and Whiskey rounded the corner at a smart pace. Unused to encountering cars like that he skidded to a halt and glared at them. He was huffing and puffing a bit because I'd hurried him on so poor Dick and Sue thought they were confronting a snorting bull at least. Then I reappeared and hustled Whiskey toward his own field. I managed to puff out,

'Only be a minute now,' and then, because I felt a bit guilty about their poor welcome, I added 'Lovely to see you,' and disappeared again. When at last I returned they waited to see if I was staying before speaking.

'Er, do you do that often?' asked Dick.

''Fraid so,' I admitted. 'It's Whiskey that's the trouble.' I saw Dick exchange a knowing look with Sue so I hastily added, 'The bullock's called Whiskey.' At long last I managed to usher them both into the house and keep Topper out, quite an achievement, but by then both dogs were barking their heads off. When Sue could make herself heard she said, 'How do you put up with all these animals, I couldn't cope.'

'I love 'em all really and you soon get used to them.'

'Sue wouldn't,' Dick grinned. 'She won't have a dog or a cat at home.'

'I like to keep the furniture clean and I couldn't bear animals in the kitchen,' she said to me. I furtively glanced around and saw to my relief that no cat marred the place, they must have taken offence when I cleared them out earlier. I opened the back door invitingly and watched the two dogs amble out to the outside room. They didn't mind, they could sense they were not welcome.

'You're certainly out in the sticks here,' Dick said chattily. 'Very different from your old house, bet you miss it?'

'Oh yes,' I told him but not in the way he meant it. 'It's lovely here, so peaceful and half the time I don't know what day it is.' At that moment one of the cockerels decided to let the world know that he was the greatest and Ted revved up the tractor in the yard. Things were not going to plan. We continued to make conversation but it was clear that my new way of life was an incredible nightmare to Sue and Dick and they couldn't understand what the attractions were.

'Want to see a piglet?' shouted Ted from the kitchen door. Normally Sue would have crooned over a baby animal but this one was cold and dirty and smelly. It was a bit weak and had been pushed away from the heater in its pen. I washed it off in the sink and dried it in the dog's towel but it obviously had no appeal as far as Sue was concerned.

'What on earth do you do with it now?' She had a look of distaste on her face.

'I usually pop it in the low oven,' I faltered. 'Just for a little while,' I added. 'And not when I'm cooking in it.'

'What is for lunch, then?' Dick meant to be funny but it misfired as I replied, 'Pork, one of our own pigs.'

The visit was not a social success and I wasn't sorry when we stood outside to see them off. The afternoon had turned misty and the moor, what you could see of it, looked dull and miserable. I heard a drumming noise coming nearer and guessed it was the last of the local hunt going home after their day out. Dick and Sue stood mesmerized as the drumming changed to a steadier clip-clop, the riders were off the grass of the moor and on to the road. They emerged from the mist as Sue whispered, 'Whatever's that?'

The seven riders passed us, touching their hats with their riding crops. One of them I knew quite well. He paused to speak and his mount stretched toward Sue hopefully because she had a bag of carrots I'd given her. She squeaked in alarm and backed away. The rider pulled his horse away and spoke over his shoulder to me.

'We've driven all the foxes up this way, I know you've got plenty of poultry for them.' I laughed, he thought our free range birds a bit of a joke. 'Thank you,' I retorted. 'I'll call a curfew tonight.' I turned back to Sue and Dick who were in the car.

'I honestly expected to see a headless horseman then,' Sue said with a weak smile. 'This mist is so spooky, I'd be scared to live here.' We made our farewells and they were gone, back to civilization I thought. I stood for a while in the mist. It was quiet but I could hear a hen cackling, the pigs had been fed so there was only the odd grunt coming from the pig houses. The geese made their stately way across the yard, then Topper with Peggy on board came into view. Peggy was swaying gently on Topper's back, they briefly acknowledged me as they passed. Topper paused to open the gate with her head and made her way to the back door where she waited hopefully. They knew I'd be going in that way and that they might get a tit-bit. They got one. When Ted came in he asked me if I wished I was going back to our old house.

'Not a bit,' I replied truthfully. 'Dick and Sue think we're potty, but I'd rather be potty here than so-called sane there. Sue reckons it's uncivilized here but . . .' I spoke rather indignantly as I took the much recovered little piglet out of the oven '. . . I don't think we're uncivilized, do you?'

It isn't often that something unexpected doesn't happen in my day. After the visit from Dick and Sue, Ted went off shopping and returned with an electric fencer.

'I'll go and get the posts in,' he sounded eager. 'Then perhaps you'll help me with the wiring. Won't take long with the two of us,' he promised. It took all day. That's the sort of thing that crops up.

Quite uncharacteristically I made up my mind to enter a cake in a competition at the local flower show. I had, originally, meant to enter a collage of an owl made of split peas, lentils and grains of rice stuck on a card. Foolishly, I'd left it on the kitchen table for the glue to dry and the window was ajar. By the time I came to wrap it up Peggy was in and busily eating my collage. She enjoyed the lentils and the rice and was on the last of the split peas. Sadly I regarded the bare piece of card and thought to try my hand at cooking instead. The next morning I'd assembled the ingredients and had got it ready to pop in the oven when Ted called me.

'Can you just stand in the gateway for a minute, only Joe's bringing a bunch of bullocks along to let out on the moor and we don't want them getting in with ours, do we?'

We had our electric fence now and the single strand of

wire across the gateway was enough to keep our cattle in, but Joe's didn't know an electric wire from an ordinary one and might push through it before they realized the drawbacks of such an action. Ted, in any case, was gone before I could answer. I shoved the cake in the oven, sadly regarded the mug of coffee I'd just made and went to stand in the gateway. It was clouding over and looking like rain. Perhaps I should have slipped on a jacket but Ted had said to stand in the gateway for only a minute so the bullocks must be coming along soon.

I waited, expecting to hear them coming. It remained very quiet. Peggy came to join me and she scuffled about in the grass at my feet. Topper came along and looked at me enquiringly. The first drops of rain pattered down and goat and chicken took themselves off. Fair weather friends only, I thought. Surely the bullocks would be here any minute. I daren't pop in for a coat. I continued to stand there while one of those showers that saturate you in seconds passed over me. Still no bullocks. I stomped angrily to the corner. What was Ted playing at? I rounded the bend (in more ways than one) and now I could see Joe's barn. In it, sheltering from the rain, were Ted and Joe comfortably sitting on a couple of hay bales.

'Morning,' called Joe.

'Ah, there you are,' said Ted. 'We're sheltering from the rain. I thought you'd guess we wouldn't start until the rain stopped, I know you don't like getting wet.'

Cooking with animals around can be a trying experience: I try to cook it, they try to eat it. It becomes a battle of wits. I'd intended to make a beef casserole for a couple of girlfriends. In plenty of time the night before I'd got the beef out of the deep freeze and left it in the larder to thaw. By evening I realized it wasn't going to thaw as fast as I wanted it to. Last thing I made quite sure that no cats or dogs were lurking in the kitchen, and I left the meat in there. All under control. What I didn't know was that Ted got up to a farrowing sow in the night and out of the kindness of his heart let Buttons the black cat in, on account of it raining . . . Now, a lot of recipes allow for economy of meat in a casserole, but not for the total lack of it. I couldn't use anything from the freezer at this late stage so I had to invent a vegetable casserole and serve it with sausages. It turned out very tasty and we all enjoyed it. Buttons did not get fed again that day.

The time I decided to follow a friend's recipe for a Christmas cake was not without incident either. The culprit was Polly, a long-haired tabby with big ideas about herself, a dignified cat who cannot bear to be laughed at. According to the recipe I had to soak the raisins, sultanas and currants in brandy overnight. This I put ready in a dish covered with cooking-foil. To be sure that nothing would happen to it I placed the dish on a top shelf in the kitchen.

I went to get some supper for Ted and me later on and noticed to my horror that Polly's tail was dangling from the shelf. On closer inspection I found the rest of Polly lying curled up beside the dish of fruit and brandy – well, some of it. There were shreds of cooking-foil around her and Polly was fast asleep. In response to me shouting at her she did try to sit up, but the method had escaped her memory temporarily. She flopped sideways which panicked us both and I grabbed at the dish. She thought I was grabbing at her so she lunged along the shelf and down on to the working top below. Taking a very zig-zag course she made her way to the edge of the working surface and without any hesitation she walked off into space, hit the swing top of the rubbish bin and disappeared into it. I lifted off the lid and Polly's eyes glowed back at me from the darkness within, I could just make out an eggshell sticking to her fur. I was shaking with laughter as I hauled her out. I hadn't the heart to be cross, she looked so indignant. I placed her outside and watched her stagger across the yard. Halfway she met Buttons who greeted her in his usual way by sniffing at her face. He recoiled in disbelief and he, too, saw her travel sideways into the barn. Polly slept there for a full twelve hours and then woke up with a raging thirst. I know, I saw her lapping furiously at a puddle.

Whenever I go to the larder there is often the full contingent of three cats and two dogs all waiting to see what I'm getting out. Because of the attentive audience who frequently try to get in the larder I tend to take out what I want as fast as possible and shut the door quickly. This was why I mistook chocolate blancmange for gravy once and nearly spoiled our meal.

Of course it would be easier to ban the animals from the house but I haven't the heart, so life continues with its rich patchwork of peaceful times and the other sort which I've just described.

Chapter 10

The Village Pub and Social Life

I am not a great drinker although Ted loves his pint or two (where do men put pint after pint? I struggle for ages with a small Guinness) but I do enjoy going to the Fox and Goose. It is a mine of information. I thoroughly recommend anyone living in the country for the first time to patronize their local. We found where to get cheap timber, pet food and firewood. I learnt to make brawn and we joined in the barter system; in our case we offered excess eggs and garden manure and in return we got fresh vegetables, fruit and home-made wines. We offered to do the odd transport job with the Landrover and we got help when building our pig houses.

Soon after we arrived in the area, Sid the landlord got a pool table in the bar. Ted had played before so he was happy to have a game with anyone. I got bored watching so with much trepidation I tried my hand at a game. After making a fool of myself on frequent occasions I actually beat someone. From then on I became keen and had another reason for enjoying my visits to the pub. When my mother heard about my new social interest she was a bit perturbed.

'Is it quite the sort of game for a woman your age?' she enquired. 'Are you happy about Liz playing?' she asked Ted.

'Lord yes,' he grinned. 'But I did tell her not to wear her split skirt and low-cut satin blouse, I thought that was a bit much.' Poor Mother!

Sid arranged a pool knock-out competition to be held one evening, each player to pay a small sum for charity and the winner to get a trophy. A lot of support was forthcoming and it became obvious that to fit it in to one evening we'd have to play in pairs. There weren't enough of us females to 'go round' and make up mixed pairs so it was facetiously suggested that some of the men should dress up in drag. Amazingly the idea caught on; although no-one admitted to approving, no-one refused. Where two men were playing in pairs a draw was made to decide which one should be the 'girl'.

On the appointed evening Ted and I walked into the bar and noticed the back view of a slim blonde on one of the stools. She swivelled round as we approached and to our utter delight we (only just) recognized Sid the landlord in a tight skirt, black tights and a figure-hugging sweater. All this was set off to a nicety with a curly blonde wig that contrasted horrendously with his own black beard.

As the evening wore on and the place filled with some fascinating sights the conversation got intriguing.

'If I says I don't half fancy you, Sid, do I still have to pay for my beer?'

''Ere, is that one of them living bras you got on under there, only I just seen one of 'em move.'

'What you got your bust made of? I tried a couple of dusters but they looked all lumpy so my sister gave me some cotton wool.'

'Mine's a couple of them thrown away nappies, they're lovely and soft, feel 'em.'

'Not bloody likely, mate, what d'you think I am?'

One of the older players, who could have a been forgiven for refusing to dress up on the grounds of maturity, turned up in his wife's long evening skirt and a long sweater that hugged his rather portly figure so well that his friend was moved to say, 'Best boil a kettle, Henry here looks ready to give birth any minute.'

'I quite agree,' was the rejoinder. 'But I reckon it looks likely to be twins. Do twins run in your family, Henry?'

'Run in his family? I reckon they'd gallop if they thought Henry was to be their mother.'

'Very witty,' said Henry seriously. 'Now if you party have had your fun, one of you can hold this handbag for me while I get on with this 'ere game of pool.'

'Tell you what, Henry, your boosooms is committing a foul on the pool table, your left one just moved the yellow ball.' Henry straighted up majestically from the pool table.

'As you're not the referee, George, you ought to be ashamed to admit that you were looking so closely at my boosooms at all.'

One of the regulars, a tall handsome fellow with a bandit-style moustache, had turned up in a smart little black dress which looked a lot more little on him than on his sister. He sported a wide-brimmed hat with a cabbage rose and a lacy shawl which failed to cover his hairy arms. The best sight of all was to be seen at floor level. He had on his normal footwear: cowboy boots! This eye-catching figure struggled manfully with the shawl until his cue became tangled in it, but he persevered with the hat until the last game, by which time it had slipped down over his eyes and he was peering out from under it to see the pool table. He also experienced some difficulty in bending over in his tight dress.

'Mind how you bend over, lad, you're showing a lot of leg,' someone said.

'He's showing a lot of cheek from where I'm sitting.'

'D'you know,' Henry enunciated his words carefully, 'the way he's peering out from under that daft hat of his he reminds me of my ferret – they both got pink eyes.'

In the car park after a most unusual evening, Ted and I watched delightedly as the moustached 'lady' got into a romantic embrace with his girl-friend. They both had tight skirts but only one wore cowboy boots and a hat that kept slipping between their kiss.

On occasions Sally and I have reason to dislike the Fox and Goose, which is unfair because it's not Sid or the place that's at fault, it's the people who go there, like Ted and John. For instance: Ted had gone into town to collect some stuff from the vet and had promised to get a couple of sliced loaves for me to make sandwiches for the WI Xmas party that evening. He assured me he wouldn't be long so I had got the fillings already waiting – and waiting. The phone rang.

'Liz? I'm at the Fox and Goose. I met John in town and we popped in for a pint, only Jack-the-Wine is here and he wants to give us a bottle of his home-made wine for Christmas so we'll nip into his house and fetch it, but I won't stay so I won't be long, OK?'

He rang off before I could say anything and my heart dropped. I hadn't been able to ask if he'd remembered the bread, also Jack-the-wine, as his name implied, made a lot of wine and he was noted for 'tasting sessions' at his house. Oh dear, I sighed.

Two hours later I heard the car come into the yard, Ted entered the kitchen, his face aglow with a silly grin. He leaned against the door for support.

'You'll never guess where I've been, never ever, never ever.' He started to laugh quietly to himself.

'Oh no?' I asked grimly.

'I've been to Jack's house wasting time, no, tasting wine with my friend John.' He began to slide down the door, realized what was happening and hastily straightened up. He began to search for a cigarette, feeling all his pockets except the top one where his cigarettes were sticking out.

'Have you got the bread?'

'Bread?' he repeated blankly.

'You said you'd get some,' I snapped.

'It's not snowing yet, we shan't need any. Plenty in the freezer.'

'I wanted sliced, not frozen stuff.'

'Sliced, ah yes, jolly clever whoever thought of it.'

'Have you got any or are you too tight to know?' I was getting desperate.

'Tight, yes I am, Liz, as a newt.' He found the cigarette packet at last and fumbled with it, trying to get one out. He put the wrong end in his mouth. He found some matches on the kitchen window-sill. In spite of my annoyance I watched fascinated as he opened the box the wrong way up and scattered the matches on the floor. I thought it might hurry things up if I retrieved them.

'For goodness sake, let me light it for you,' I said angrily, hoping that once it was lit he'd move away from the door and I could look in the car for the bread.

'You've got it the wrong way round,' I told him, meaning the cigarette. Ted obediently turned the matchbox over in his hand and attempted to strike one on the smooth side. I removed the cigarette and replaced it and lit it. At that moment there was a banging on the door and I pulled Ted away as it was opened and John's face came into view.

'I say,' he was indignant. 'You asked me in for coffee.'

'You were asleep when we got here,' Ted replied and he moved to sit at the kitchen table, putting his hand in the grated cheese I'd left ready for the ill-fated sandwiches.

'I was only resting my eyes,' John protested as he stumbled across the room. I could see that focusing wasn't his strong point either.

'We went to Jack's together,' Ted began conversationally.

'Yes, we went to Jack's together,' echoed John.

'He insisted we tasted his wines,' added Ted.

'Tasted his wines,' John echoed again.

'Everyone of them I should think,' I snapped.

'Most of them,' Ted owned up, looking at his cigarette as if he'd never seen one before.

'Yes, most of them,' said the echo.

'I didn't think the beetroot was too good,' Ted sounded thoughtful.

'No,' agreed John. 'I didn't like the beetroot at all.'

'You drank enough of it.' Ted sounded surprised.

'Not me, you did.'

'I didn't.'

'You drank two yogurts spots full – two yogurt spot full—'

'Yogurt pots?' I queried.

'That's right,' said John.

'Why yogurt pots?' I persevered.

'Didn't like to refuse,' Ted explained. I gave up.

'Did you get any bread?' I tried.

'No, not bread,' John assured me. 'There was gooseberry and apple and—'

'Bread,' I repeated. 'Sliced bread?'

'No thanks, Liz, I'm not hungry.'

I plonked two mugs of black coffee in front of them and went outside to the car. On the back seat were two sliced loaves, four bottles of wine and an empty yogurt pot. I felt much less snappy and ushered the two of them into the front room while I did the sandwiches. I peeped in later and found a peaceful scene. Ted was on the settee with the cat on his chest, sound asleep. They were both breathing steadily, the cat rose and fell with each breath John was on the floor, his arm over the old dog's neck. I rang Sally.

'Hello,' she said. 'I was just going to phone you, is John there?'

'He is,' I told her.

'What's he doing?'

'Sleeping with Ted.'

'Doing what?'

'They are sleeping it off,' I explained.

'Sleeping what off?'

'A visit to Jack-the-wine.'

'Oh no. Oh well, I suppose it had to happen sometime, it's been threatening hasn't it? Had any trouble with them?'

'No, it was quite funny really.' I told her all about it.

'Tell you what, Liz, we'll go to WI together and have a drink after at the Fox and Goose.'

'Jolly good idea, we'll have a game of pool, too.'

'Lovely,' she agreed.

Sometime after that I became involved in one of John and Sally's marital disputes. Ted and I enjoy their company very much but their relationship is emotionally volatile and a spark can start an explosion. It's usually over as quickly as it begins but at the time it's dramatic. Our phone rings and I find I have Sally in tears.

'Liz, it's awful. John and I have had a row, it's all finished between us, it's over. He's left me.'

'He's what? Where's he gone?' I nearly add, this time, but think better of it.

'I don't know where he's gone,' wails Sally. 'He took the car and went off.'

'I shouldn't worry too much, he'll be back.' I try to calm her.

'I don't care if he never comes back. He's beastly,' she snaps.

'Well, there you are then.'

'He might have an accident, he was in a temper,' Sally wails again.

My own private view is that John might easily have an accident any time, the way he drives.

'And there's only a little bit of petrol in the car and he doesn't know that so he might get on to the motorway and run out and have an accident.' At that moment, John drives into our yard at about ninety miles an hour, skids to a halt and gets out smiling broadly.

'I don't think John will get on to a motorway,' I say slowly.

'Well, I suppose he might have gone to the Fox and Goose.' Sally sounds a bit calmer. He'd have had enough petrol for that.'

'That's right,' I say reassuringly.

'But he hasn't any money with him, that's what we quarrelled over. I was to have got some from the bank this morning.'

'And you forgot?' I hazarded.

'No, I got it but I spent it.'

'Oh, Sally, you didn't?'

'There was this auction sale on and I just popped in for a look but I saw this super chess set and I – er – but that shouldn't have made him cross. I only spent the housekeeping and I could have fed us this week on what's in the deep freeze.' I sighed, the last I'd seen in Sally's freezer it had contained a bag of blackberries and a loaf of bread. 'John wanted to borrow from the housekeeping because he forgot to get his money. He was furious, he drives fast when he's cross and round these lanes—' She trailed off, sounding truly anxious. I decided to be honest.

'Stop worrying, Sally, I know where he is. He's just arrived in the yard.'

'What?' Sally shrieks down the phone. 'He's there?'

'He'll probably have a chat with Ted and be quite all right after.'

'You tell him from me that I never want to see him again and I'm giving his dinner to the dog now.' She slams the phone down. I trail to the kitchen where Ted is washing his hands and John is trying to persuade him to take us all to the village for a can of petrol and perhaps a pint? I decide to go with them because I can do with a few things from the shop. As Ted and John settle to their second pint I ring Sally from the box outside the pub. She answers as bright as a button.

'Liz? Guess what? I think that chess set might be ivory. I'm just cleaning it, isn't that super?'

'Lovely, but I thought you were worried about John,' and I tell her what's happening. 'I don't think they'll be long.'

'Tell you what, I'll come down on the bike and join you.' She was gone.

I puzzle over this idea because the only bike I've seen is

certainly not roadworthy. I wait a bit anxiously for developments.

Instead of rejoining Ted and John I hovered outside the pub. Sally eventually swooped into the car park, skidded to a halt and fell off with the bike on top of her.

'Oh my goodness, Sally, are you all right?' I queried as I lifted the bike off.

'Fine,' grinned Sally, as she brushed the dirt off her jeans. 'It's the only way to stop on this, no brakes you see?' John came rushing out.

'Whatever did you ride that old thing for, you could have hurt yourself.'

'I didn't think you'd care if I had,' retorted Sally.

'Don't be silly, of course I'd care.'

'Would you really?'

'I've just said so, haven't I?'

'Yes, but after what you said to me earlier—'

'That was earlier,' John put his arm round her and I wandered back to Ted as I appeared superfluous.

'I think they're making it up,' I told him.

'Making what up?'

'Their argument, didn't John tell you?'

'No, never mentioned it, had another one did they?'

'Yes and now they're making it up.'

'Oh well, that's all right then. Want a drink?'

That sort of incident isn't so bad but one happened when Cliff was with us and he wasn't prepared, well, nor was I, particularly as it took place at gone eleven at night.

We'd had a hard day moving pigs from one house to another so Ted and I went up to bed early. Cliff always sat in front of the television, no matter what, until the dot disappeared. This evening he was half-asleep when he was startled by Sally bursting through the door in tears. He leapt from the chair not sure whether he was watching a TV drama or the real thing. Cliff had only met Sally briefly once.

'He's turned me out and locked the door and I've nowhere to go,' blurted Sally before she realized that neither Ted nor I were in the room.

'Er, who are you?' faltered Cliff.

Afterwards we had a good giggle at it all, but at the time Cliff was shattered. He stumbled up to our room and knocked on the door.

'Liz, can you come downstairs, I think there's someone to see you.'

'Who?' I asked.

'I'm not sure,' he answered plaintively. 'But I'll wait until you're ready before I go down again if you don't mind.'

Gentlemanly as always, Cliff waited until I had my dressing-gown on and followed me to where Sally was hunched up in an armchair. Once he realized it was a domestic dispute and not an attempted burglary he scuttled up to his bed.

'I'm ever so sorry to disturb you,' sobbed Sally. 'I didn't know it was so late.'

'Well, it is rather and we've been busy,' I began, then I felt ashamed. It must be serious for her to arrive at night.

'It's John,' she told me unnecessarily. 'We had a quarrel and I said I wasn't staying another minute, then he laughed and said where on earth was I going to go at night in the middle of the moor, so I got furious and said I'd take the car but when I got in the keys weren't there and by then John had locked the house door and he shouted at me to stay in the garage until I saw sense so that's why I'm here,' she finished, blowing her nose violently.

'How did you get here?'

'I walked.'

'In the dark, across the moor?' I was incredulous, Sally hated the dark.

'I was furious,' she said by way of an explanation.

'It's nearly two miles!' I exclaimed.

'I couldn't have stayed, could I?'

'You shouldn't have threatened to take the car, Sally, you haven't even passed your test yet.'

'I'd only have driven here.'

'Even so, you told me yourself you couldn't turn it round and I bet you don't know where the light switches are or anything.'

'I suppose it was a bit silly, but I was so cross.'

'What about?' I sighed and sat down.

'I want a baby and John doesn't.'

'A baby?' I was astonished, Sally had never mentioned a family. Then I thought I understood.

'A baby what?' Sally loves animals and has kept guinea pigs, mice, cats and dogs, of course, and even looked after a boa constrictor when its owner went abroad for six weeks. Last thing she'd mentioned was breeding llamas.

'A baby baby,' she snapped irritably. 'I want one before I get too old.'

'I thought you said you wanted to go back to university for a year?'

'Yes, yes but I'd rather have a baby.'

'And John doesn't want one?'

'No he doesn't and he won't even talk about it.'

'Didn't you ever discuss it before you got married?'

'Oh yes, I didn't want one then but I've changed my mind.'

'Quite a change,' I said gently. 'When did you have this change of heart?'

'When I visited my sister and saw hers and started thinking about it—'

'Sally, Sally, that was only last weekend! Did you jump this on John out of the blue tonight?'

'Er, well, I suppose it was a bit sudden but he didn't have to be so horrid and say that I couldn't even look after myself let alone a baby.'

'Look, he's probably suffering from shock and you do tend to get yourself in some fixes, locking yourself out of the house then climbing a ladder to get through the window and falling off.'

'It slipped.'

'I know, but it was only just after that you fell in the river and had to be brought home from the village soaking wet.'

'The dog couldn't get out.'

'It got out quite happily all by itself when you fell in.'

'But—'

'And you gave that tramp John's jacket.'

'The poor chap was cold!'

'He then took John's tools from the garage and sold them to Sid at the Fox and Goose.'

'I thought you'd understand.' Sally sounded sulky.

'I do, but you must try to understand John's feelings as well.'

'I suppose,' she sniffed reluctantly.

'I'll run you home,' I offered.

'I can't face him tonight,' she muttered. In a way I was relieved.

'Stay in the spare room then, but phone John and let him know where you are.'

'No, let him worry,' she said stubbornly. I saw her into bed with a hot-water bottle and sneaked down to phone John. He took ages to answer.

'Sally's here,' I told him.

'Thought she would be,' he answered sleepily. 'I'd gone to bed.' My sympathy for him evaporated and I went to bed too.

'What's going on?' moaned Ted.

'John and Sally again.'

'I heard her voice, that's why I stayed up here.'

'Coward,' I said.

'Too right, where is she now?'

'In the spare bed.'

'Oh Gawd,' and he covered his head with the bedclothes.

'Men!' I muttered and deliberately put my cold feet

against his legs.

In the morning, I found Sally already making tea. She was dressed and very apologetic for disturbing us, especially Cliff.

'I didn't think of anyone else being here,' she said. 'He looked ever so surprised to see me.'

'I'm sure he was,' I agreed and wasn't amazed that neither Ted nor Cliff ventured out of bed until they heard Sally leave. She insisted on walking home to make John sorry. I had my doubts about that but thought they'd soon sort things out as they normally did. I was wrong. Sally reappeared in the afternoon, steely calm and carrying a haversack of belongings.

'I would like to stay in your caravan for a couple of days please, Liz.'

'Oh come on, things can't be that bad.' I was afraid they were, though, I hadn't seen Sally like this before. Tears and temper, yes, cool and calm, no.

'Only until the weekend and then I'll go to my parents, at the moment they're on holiday.'

'Sally, I—'

'I'll pay you for the use of it,' Sally stated quietly.

'For goodness sake!' I exploded. 'You can't do that.'

'I'll either stay here or go to the Fox and Goose.' In the face of that argument I led her to our holiday overflow caravan. We don't let it, but it is used as a spare bedroom when friends arrive all at once. If she went to the pub the whole village would know about their troubles in no time. The arrangement was uncomfortable. Sally was prone to telling Ted, Cliff or me yet another reason for her leaving John. We heard nothing from John but after forty-eight hours of Ted and Cliff skulking round corners there was a phone call for her. It was John. Sally came out of the room tight-lipped.

'John is coming to collect me to take me to my mother's. I shall go and pack.' I didn't know what to say. I was used to their tiffs but not this sort of thing. I offered to help but she refused, merely thanking me for having her. I felt terrible. John arrived and I watched them depart. The three of us felt depressed. It started to rain and we sat miserably in front of a television that we didn't want to watch.

'Tell you what,' said Ted at last. 'We'll feed the pigs a bit early tonight and go down to the Fox and Goose and have a meal out. How's that?'

'I didn't know they did meals?' Cliff was surprised.

'They've turned the back room into a restaurant, it's only just opened so we'll christen it for them.'

'A super idea,' I was relieved. 'Take our minds off other people's troubles. After all, we can't help them by sitting here,' I pointed out to Cliff who was looking a bit worried.

'It's not that, Liz, but you know I don't like funny foods and foreign stuff to eat.'

'It's not like that,' Ted assured him. 'You don't have to eat snails if you don't want to.'

'Thank goodness for that,' said Cliff who was very plain in his tastes.

'No, you can skip the snails but I hear the frogs' legs are good.'

I phoned to book a table as I knew Sid was hoping people would, at least until they got into the swing of it.

'Sid? Can you manage a table for the three of us this evening, it's the works outing.'

'Well, I'm blessed.' Sid sounded surprised. 'I've just had Sally on the phone and she's booked a table for two.'

'What? For two, her and John?'

'That's what she said. By way of reconciliation, I think they've had one of their quarrels again.'

'Yes, they have. We were caught in the crossfire this time and we thought they'd really split up. We were coming down to cheer ourselves up.' Sid gave a chuckle.

'Best leave your meal to another evening, I reckon?'

'Better,' I agreed. 'Don't think I could face it just now.

But we will come, Sid, I promise.' I gave the news to Ted and Cliff. Their faces were a study.

'D'you mean,' Cliff was bemused, 'that after all that, they've made it up?'

'Sounds like it.'

'If we go somewhere else for a meal tonight, we'll still have to go to Sid's sometime to be fair to him, so John and Sally owe us a meal out.' Ted decided.

'And they say living in the country is peaceful,' Cliff moaned. 'It's like the back page of a woman's magazine.'

'Never mind,' I brightened up. 'That's two meals I'm not going to have to prepare. Come on let's get ready, I'm hungry.'

'You women are heartless,' Cliff remarked. 'Making a profit out of other people's troubles.'

'I've earned it,' I replied. 'I didn't scuttle up to bed in the face of danger like you did, nor did I hide in bed when Sally turned up.'

'OK, OK.' Ted leapt out of his chair. 'Let's go before anything starts up here.'

We had a lovely evening out for which Ted threatened to send the bill to John but we didn't see them for a week or two. By the time we did meet again, everything in the garden was lovely so we didn't refer to the quarrel in case of stirring things up. Cliff maintained that he'd suffered the greatest shock when Sally burst in on him.

'I thought it was an action replay from Dallas or something.'

Chapter 11

Every Cloud Has One

We were getting nicely into the swing of pig farming, sending off good healthy pigs and getting a cheque back in due course, when what could have been a disaster struck. It was so simply done: Ted got off the tractor and his foot slipped in a sticky patch of mud. He didn't fall but he twisted round. At the time he hardly took any notice of it but when he'd come in and sat down at the table for elevenses he gasped as he stood up.

'Hell!' he exclaimed, clutching at his back. 'That hurts, I can hardly move.'

'Sit down again,' I suggested.

'Can't, not for a minute.' He leaned on the table for a bit, then painfully stood. 'Queer that,' he said. 'I hardly felt anything when I got off the tractor but that's when I must have twisted it.' He began to walk to the door and the cigarettes in his top pocket fell on the floor. Without thinking he went to bend down to pick them up. He couldn't. Standing still and upright was OK. Attempting to sit was painful, and he could only sort of perch on the edge of a chair with his legs straight out in front.

'It'll go in a minute,' he insisted, 'like lumbago.' I was worried. I could see he was in pain and I'd heard enough about slipped discs to think this might be the trouble. If so, they didn't just go.

That evening Ted hobbled painfully to the pig houses, telling me how much to feed what. I did the best I could but some of the cleaning was heavy work. Ted couldn't have got into a bath to try a bit of heat treatment so I placed hot-water bottles at his back. Eventually he found a position that wasn't too uncomfortable and managed to get a bit of sleep.

In the morning he couldn't get out of bed. In the end with his legs to the side and me heaving gently we got him upright, sideways. He had to hang on to my shoulders to get on his feet. No way could he do any work. To get him to the doctors was a problem. To start with, Ted classes all medics as 'quacks' and didn't want to go. I insisted, pointing out that I couldn't see to the pigs properly for more than a day or two before jobs would crop up that were physically impossible for me. After many trials and tribulations he managed to sit sideways in the back of the car and arrived at the surgery. I waited for him to emerge.

'What did he say?' was my first question.

'You'll never credit it, Liz, but after he asked what my job was, he said "Does it involve heavy work?" I ask you, what does he think. I told you it was no good coming to a quack.'

My heart sank. It wasn't the doctor's fault, after all some farmers had workers to do the heavy jobs, but we couldn't afford it.

'Is it a slipped disc?'

'He doesn't think so, he's given me some tablets. They'll do a lot, I don't think.'

The next two days were hard, to say the least. Ted felt no better and understandably was bad-tempered. He could tell me how he did things but I wasn't always equal to the task. Heavy buckets of pig nuts I could carry but I just could not fling them over the dividing walls of the pens, so the feed fell in the dunging area. I couldn't drive our old tractor to dispose of the slurry so the slurry pits were dangerously full to overflowing. It seemed to me that I had no sooner struggled with the morning feeding session than it was time to start again. Our neighbours helped as much as they could but Joe had his own work to do. Finally, Ted and I agreed we had to get someone.

'Cliff would be a help but he can't do the heavy work on his own. He's all right to help me but not by himself, but he does know what to do and where things are. You and he could help each other. I don't want you cracking up,' Ted said.

'Perhaps someone in the village could come up for feeding times and help?'

'I can't think of anyone but if Cliff could come again he can manage the tractor and could spread a bit of slurry.'

I knew Ted was desperately worried, so was I. Luckily Cliff had recently had a medical check-up and had been pronounced as well as could be expected. He was concerned but happy to come.

'Only too glad to be of help, you know I'll do all I can,' he said on the phone that evening. 'I'll come by coach the day after tomorrow.'

I then started to worry that he might try to do too much and make himself ill so I rang Sid at the Fox and Goose. I told him my tale of woe.

'Now look here, Liz, why didn't you tell someone before? I think I know of a young lad who's left school and hasn't got a job who'd enjoy working at your place. He wouldn't want much in the way of wages so you stop worrying and I'll get hold of him for you as soon as I can.'

Ted was not optimistic. 'I don't want some teenager full of pop music and dreams of girls,' he grumbled.

The next morning a spaceman arrived on a motorbike. As the helmeted figure dismounted I saw he was wearing overalls and wellies. At least he looked businesslike, I was relieved to see. From under the helmet emerged a cheerfully grinning face.

'Morning,' he greeted me as he unwound a mile-long scarf. 'I'm Keith. Sid tells me you got a bit of trouble. I've worked on my uncle's farm and he's got pigs.'

'Am I glad to see you, come in and meet my husband. He's a bit unhappy at the moment but he isn't usually so –

er—'

'Tetchy?' suggested Keith. 'Like my dad when he's off-colour.'

'Probably,' I agreed and hoped he'd stay, I'd taken to him already.

Ted greeted him rather shortly, he was so cross at being unable to work and was taking it out on all and sundry.

'Hello,' he grunted at Keith. 'I can't get up, bloody back trouble, you can see the fix we're in and I may as well tell you that we can't afford to pay much.' What a welcome, I thought, and my spirits sank. It didn't put Keith off, he smiled understandingly.

'Just like me dad,' he murmured to me. 'See here,' he said to Ted, 'I've not been able to find a job yet and I've only just left school so I don't get dole money but I've been doing a bit here and there and I'm willing to work for what you think I'm worth. I been cleaning windows in the village and I think I've talked to enough women to last me a bit.' This final remark brought a sympathetic look from Ted and I could tell he appreciated the lad's attitude. I left them to talk over the situation and made a cup of tea for Keith. When I handed it to him he thanked me and asked if he could put a drop of cold in because he thought he'd better get started a bit sharp. Ted was looking so much happier by the minute that I could have hugged Keith there and then. The three of us did the rounds in the pig houses and I could tell that Sid had chosen well for us. Keith asked intelligent questions and did what was required at the same time. As we came back to the house, Ted looked tired and drawn with the effort of standing and talking. I suggested he had a lie-down for a bit. Keith and I had a talk on our own. He agreed to have lunch with us but as part of his pay.

'Would it be all right if I started about eight in the morning?'

'Fine by me,' I told him. 'Then if you do the afternoon feed about three, you can be away by four.'

'Smashing,' was Keith's reaction. The afternoon feed is quite quick and the two of us worked together well. Keith had begged a bit of paper and pencil to make a few notes and I felt he knew as much as me already. Ted came down from his lie-in looking only a little better but he said cheerio to Keith with a grin. He admitted that he seemed a good lad and he was glad he'd come. That night, Ted slept badly so I left him when I got up as he was dozing. I glanced a bit anxiously at the clock, noticing it was nearly half-past eight when Keith came to the door.

'I'm already here,' he announced. 'I've done the weaners and the growers, thought I'd better tell you in case you were thinking I was late.'

'Keith, you're an angel,' I told him. 'Any problems?'

'Don't think so, I jotted down their amount and I've cleaned them out.'

'Here, drink this cuppa and then I'll come down while you do the sows and I'll do the farrowing house. It'll be lovely not to have to do it all.'

'How's the boss?' he enquired.

'Bad night but he's sleeping now.'

'Let's see if I can have finished before he wakes, then.' Feeling like happy conspirators, we went to finish the work. I had started to cook breakfast when there was a bull-like roar from above.

'Reckon the boss is awake,' remarked Keith.

'Liz, what's going on? It's half-past ten.' He appeared in the kitchen in his pyjamas, looking crumpled but a little better.

'I'm cooking breakfast, want some?'

'But the pigs? What about the pigs?' he sounded tetchy. Then he saw Keith. 'Oh, you're here – good lad.'

'We've done the pigs,' I announced smugly.

'What, all of them?'

'Yes, all of them. Keith has done a grand job, he'd started before I came downstairs.'

'Did you now,' Ted sounded relieved. 'Sows OK?'

'Fine, sir.' (That sir pleased Ted, I could tell.) 'Number twenty has to be moved to the farrowing house today, you said?'

'That's right, good, good.'

'Keith helped me put number twenty's pen ready, lamp and all, so we've only got to move her over.'

'So everything is done?' Ted almost sounded disappointed that he was so dispensable, but as he went to sit at the table his painful back made him realize he couldn't cope without Keith.

'How are you?' asked Keith.

'Bloody awful,' Ted replied. 'Those tablets from the quack do no good, I told you they wouldn't.'

'You could go back to bed,' I suggested.

'Just get stiff.'

'Watch telly, then?'

'Nothing on.'

'Read.'

'Read all the library books.'

'The rugby club sends blokes to an osteopath when they hurt themselves,' Keith said.

'I'm not seeing another quack,' protested Ted.

'This one does sportsmen,' persisted Keith.

'Does them?'

'Well, massage and stuff.'

'It might be an idea,' I broke in. 'You couldn't get worse.'

'Well—'

I waited for no more arguments but phoned up and made an appointment for the following day. Keith did the pigs before going home and I went to meet Cliff at the bus station. He looked anxious.

'Hello love,' he greeted me. 'I'm ever so sorry about poor old Ted. Is he any better?' I explained the situation and added that we had got Keith to help.

'You don't know how glad I am to hear that. I've been

worrying that I wouldn't be able to do enough to be much help, you know how my chest is some mornings.'

'That's OK Cliff,' I assured him. 'If you only give me moral support and take Ted's mind off things you'll be doing a grand job.'

'Difficult to live with, is he?'

'An understatement,' I told him. 'But I'm taking him to an osteopath in the morning so I'm hoping for some help then.'

'So long as I don't have to milk the damn goat,' said Cliff earnestly.

When Ted saw the osteopath he was impressed. He asked the right questions in Ted's view and gave him a 'going over'. Ted came out still in pain but as he explained, it was a different sort and he wasn't to expect much benefit immediately.

'I'm to go back next week and in the meantime I'm to take things easy and be patient. Well, I don't mind that now we've got Keith and Cliff so you don't have to do it all. Also, I have something to wait for.'

During the next few weeks, life was different to say the least. Ted was brighter because he could feel easier each day, Cliff brought his particular brand of humour into every situation and had Keith and me in stitches so often that Keith threatened to bring his tape-recorder to work. Ted was able to explain what had to be done and Keith and Cliff and I were able to tackle it. Our methods varied from the normal to the extreme but the results were pretty good. Keith stayed with us for over a month, satisfied to work for a small wage. I think the fact that we all had so much fun kept him with us, but we had mixed feelings when we were able to give him a glowing reference which helped him to get a permanent job on a farm nearby.

Just as Ted was really back to normal we had a surprise visit from Dave. He had moved into Cornwall, unbeknown to us all, and had tracked us down. Ted and Dave had been to school together and Cliff had known him some years so we had a reunion. Dave had enquired at the Fox and Goose for our farm and had learnt of our recent troubles before he even got to us. After we'd all talked together for a few hours, Dave came up with a bright idea.

'What you both need,' he said to Ted and me, 'Is a holiday.'

'I'm all right now,' said Ted. 'And we can't leave the pigs.'

'You don't need to worry about them, Cliff and I will look after them.'

'It's very good of you to think of it but we couldn't possibly,' said Ted.

'Now hang on, I'm not suggesting this lightly. Ted, you and I worked on farms together in our youth and I haven't lost touch with farming. I'm self-employed at the moment and I could stay here for a long weekend at least. Cliff can tell me what to do but if I stay here for a day or two beforehand as well there'll be no problem. You've just told me that Keith works nearby so I can always get hold of him.'

'Yes, but—'

'No buts, I can tell by the look of you both that you're tired and you can trust me with the pigs. Either that, or I'll take Liz off for the weekend.'

'I reckon I can trust you with the pigs better than with Liz, you always pinched my girls when we were young,' laughed Ted. And so it came about, the cloud of backache became silver-lined. A holiday! An idea like that had never entered our minds. Dave was a practical chap, strong and healthy. His first love was farming and I could tell he was dying to become Farmer Dave for a spell. Cliff was slightly less enthralled.

'Whose going to milk that bloody goat?' he asked suspiciously.

'I'll milk the goat,' Dave laughed. 'You can be the cook.'

'Thank God for that, I can manage a frying-pan.'

As we went to bed that night I asked Ted if he thought Dave really meant what he said.

'I rather think he did. It's the sort of thing he'd offer to do. Loves farming and I know he wants to get back into it eventually. Practical sort of chap, I'll be quite happy to let him have the pigs. Still, he may have second thoughts.'

A couple of days went by before we saw Dave again. He turned up and handed an envelope to Ted.

'Here you are,' he said. 'Get out of that.'

In the envelope were two tickets to Guernsey plus hotel bookings for three days.

'We go on December the fifth. That's the week after next,' I spluttered.

'No point in hanging about,' Dave said. 'That's when I can fit it in. You fly over on the Friday and fly back on the Monday.'

'Fly?' I whispered.

'That's right, from Plymouth airport.'

'Oh Lord.' I looked at Ted.

'What's the matter?' queried Dave.

'Nothing, really, it's just that I've never flown before.'

'Never flown? Well, it'll be a new experience for you. I thought everyone had flown.'

'I haven't either,' admitted Ted.

'Well, there's a thing, killing two birds with one stone.' Dave laughed heartily. I sincerely wished he'd put it another way.

The day before we were due to leave found me panicking. I hadn't had to pack for a holiday for years. I couldn't think what to take. How do these people manage in films? They pull out drawers and unhesitatingly throw the contents into a suitcase. They never pause to wonder if they'll need a bikini or a woolly vest. You don't see them pull out a sweater only to find there's egg down the front. I was rushing up and down to the airing cupboard like a yo-yo, trying to find a bra that wasn't grey-coloured and a pair of underpants for Ted that Topper hadn't chewed. At last, I reckoned we could risk getting run over in Guernsey.

We arrived at the airport. Nervously I smoked a cigarette and wondered why the plane I could see appeared to be taking off across a ploughed field. I wondered if I had told Cliff and Dave everything they needed to know. I'd left a long list of do's and don'ts for the cats and dogs and the poultry, not to mention Topper. If they didn't have her nuts at the ready when they called her in to be milked she'd go huffy and walk away again. Would Cliff risk looking ridiculous and stretch his arms out wide saying 'what-what-what' to the geese at dusk to get them in their house? Would they remember to—'

'Come on, Liz, stop worrying. There's our flight being put up on the board. Would you like a coffee first?'

As it happened, I spotted the bar so I asked Ted if I could have a brandy. He looked a bit surprised but got one for himself as well. We managed to look pretty cool as we got into our plane. I say 'into' because it was a small one and as I'm five foot ten inches I had to stoop to enter and walk down the aisle like a neolithic man. Once seated I felt as though I was sitting on the loo. The engines started up and I found that I was gripping my left thumb with my right hand. We started to move forward and I held my breath and concentrated on the collar of the man's shirt in front of me. It was pale blue with darker stripes and I wished like anything that I was home milking Topper. I screwed up courage to peep out of the window and I reckon we were lucky to take off at all, we only just had enough runway. Of course the stewardess looked calm enough but that's their training, isn't it?

Once my breathing had regulated a bit I quite enjoyed flying. I even had the presence of mind to accept a barley sugar before landing, well, what difference was an extra inch going to make at this stage? I could either diet later or the secret would be buried with me.

Needless to say we enjoyed the holiday enormously, we phoned home each evening but were assured that all was well. The flight home was fine, except for the moment when the stewardess informed us we were going to fly down to the eye of the wind. I thought it meant we were going into the centre of a cyclone or the core of a tornado, but Ted had read his papers and calmed me by explaining that the *Eye of the Wind* was the training ship that Prince Charles had been on, sailing. The plane went into what I classified as a steep dive. My stomach caught up at sea level. We had a beautiful view of the sailing ship but it was a very near thing with my breakfast.

It was lovely to be home again. The dogs barked hysterically, the cats wound themselves round our legs and Topper got in the car as we unloaded our luggage. Dave and Cliff went all self-conscious as we presented them with 'little somethings' and thanked them gratefully. We sat round the fire that night listening to Dave's account of things in general with wry comments from Cliff who had had all the boring unexciting chores, mainly washing-up. I couldn't stop myself from dozing off and I came to with a jerk just as Dave was saying – and in spite of all that I found her in the muck heap at the end of the field.'

I was too tired, then, to find out who or what, but as he sounded amused about it I decided it wasn't his latest girl-friend.

Over the next twenty-four hours we heard more about Cliff and Dave's adventures. The muck heap incident involved a sickly weaner. Dave had found it looking poorly the first morning and to be on the safe side he'd taken it away from the others and left it in a shed on its own. It had burrowed into some straw and Dave had left it for a bit. When he returned to it, later, it was nowhere to be seen. He and Cliff searched all over. Dave was blaming himself for having left it somewhere it could escape from, he was sure it had gone off and died somewhere. However, the morning we were due home, Dave went to empty the wheelbarrow on the muck heap and as he did so a bit of it heaved up and began to move. There was the missing animal. It was as warm as toast and appeared healthy. The warmth from the muck heap must have made it feel better, it was fine from then on.

Then we heard the 'Ballad of Cliff's Sunday' and it went like this according to Dave:

'We went to the Fox and Goose one evening, the Saturday it was, for a pint and a game of pool. What I didn't realize was that Cliff had his pension money with him and you know what he's like, it was burning a hole in his pocket and he was buying drinks for all and sundry and getting them back. He was the life and soul of the party, the court jester and all that. By closing time he didn't want to go home and was persuading Sid to give him nightcaps and whatnot. When we got back here he was full of jokes and wanting coffee so it was about one o'clock in the morning before I got him up to bed. Come the morning there wasn't a squeak out of him. I left him for a bit then I barged in and flung back the curtains and told him what a beautiful morning it was for cleaning out pigs. A white face peered out from under the bedclothes with eyes like burnt holes in a blanket. Cliff coughs and wheezes a bit and whispers that his chest is bad and he's ever so sorry but he doesn't think he can help with the pigs this morning but perhaps he'll be better to help in the evening. Mind you, I was suspicious but he did look bad so I left him while I did everything, then I came up and cooked my breakfast. I even took him up a cup of tea.

'About quarter to twelve I heard him moving around and then he was splashing about in the bathroom and I thought he must be feeling a bit better. Damn me, he burst into the room looking all spruced up in his tidy clothes and rubbing his hands together like he does saying, "Right then, are you going to get ready?"

'I guessed right away what he had in mind but I wasn't having any.

'"Ready for what?" I said, sounding innocent.

'"Lunch-time noggins, of course," he says all cheerful.

'"No way," I told him. "Not after last night and me having all the work to do this morning. You shammer, your chest wasn't bad, you were just having a lie-in."

'"Come on, Dave, it's Sunday morning. Always have a pint Sunday morning."

'"Not this Sunday morning we don't. I'm staying here."

'Well, you should have seen his face, it was a study. Nothing he could do about it if I wasn't prepared to drive him there.

'"I'm fine now, I'll be able to help this evening."

'"You may be fine now, but you were at death's door this morning."

'"A lie-in puts me right." Cliff was pleading by now.

'"All right then, come for a walk with me on the moor, that'll do you a lot more good than a stuffy bar."

'I made him come and he was just like a kid, dragging his feet and complaining we were going too far. I was dying for a pint myself by then but I wouldn't admit it. Still, it did do him good.'

'No it didn't at all,' grumbled Cliff who had been listening to all this. 'I had sore feet after.'

'Served you right,' laughed Ted. 'Playing possum like that.'

'I wasn't the only one who got up to something while you were away,' Cliff announced coolly. 'What about you and Davinia?' he said. Ted and I looked expectantly at Dave. His current girl-friend was Jennie, we thought. Dave looked a bit shaken.

'Oh, you're going to bring that up, are you?'

'Davinia?' I queried.

'You old so-and-so,' guffawed Ted. 'Still at it, are you?'

'Davinia is the new barmaid at the Fox and Goose. She started the weekend you were away,' Cliff told me. 'And she's a bit of all right.'

'Is she, indeed,' I replied.

'You ask Dave,' Cliff prompted. 'He ought to know after Saturday night.'

'What d'you know about it?' Dave was on the defensive. 'You were well away.'

'So were you – with Davinia,' teased Cliff.

'I was having a quiet chat with her. I felt a bit tired after doing all the work.'

'Not too tired to take her home. You offered her a lift and we all knew she only lives a step away in the village.'

'It was raining, I merely offered to give a young lady a lift like any gentleman would.'

'Huh!' exploded Cliff. 'You wouldn't take us both home at the same time. I had to wait for you to escort Davinia and then come back for me and you were ages.'

'You were so tight I thought it better.'

'Oh yeah,' was Cliff's parting shot.

Later on when the phone rang it was I who answered it.

'Is Dave still there?' said a feminine voice.

'He's working outside at the moment,' I explained. 'Can I take a message?'

'Would you please tell him that I can finish early tonight as long as it isn't too busy, only Dave promised to give me a lift again you see?'

I could see all right and I promised faithfully to pass on the message. Of course I chose my moment to do so, in front of Cliff and Ted. I reckoned it evened up the score.

Chapter 12

The Generation Gap

A gap is either a hole you have to fill up, as in hedges round fields, or in socks or in hungry teenagers, or it can be a way through either to fresh fields and pastures new, or a bolt-hole. I wish the phrase 'generation gap' had been dubbed 'generation exchange' instead because that conjures up a possible barter system as in 'I'll give a bit if you will,' or 'I'll mend this if you'll do that.' Mind you, I'm well aware that one can be diddled in bartering but it sharpens up the wits for next time.

With Steve and Kate living away from home I know there are a lot of things going on that I don't hear about so I don't worry, but then again I often worry about that, too, so you can't win. When I do get told things I try not to show that my blood runs cold or my hair is standing on end or I shan't hear anything another time. I do feel apprehensive when I overhear things like,

'—and her brother was so tight he slept through the rest of the party behind the settee.'

'I think she's on drugs because she seems ever so queer sometimes.'

'This latest boyfriend of Brenda's turned out to be married but they're going to share a flat anyway.'

I allow for exaggeration, but even so echoes of these confidences haunted me the weekend we thought Kate went missing. Steve arrived on his own.

'Have you heard from Kate?' he asked, 'only I thought

she was coming over this weekend.'

'No, I haven't heard,' I told him. 'What time were you supposed to pick her up?'

'The usual time, after I finish work unless she lets me know.'

'She must have arranged to do something different, I suppose, only she should have let you know. I expect she'll ring here soon and say she's off to a party or something.'

'She said a friend might be having one, only it wasn't certain.'

'That's what it must be then, she'll phone soon.' I was confident at that point. However, later in the evening when the phone did go it wasn't Kate but the friend, to say that if Kate was there would I tell her that the party was on and could she come?

'Well, where on earth is she?' I demanded of Steve and Ted.

'She'll turn up,' said Steve unconcernedly.

'Don't worry too much,' said Ted. 'You know her sense of time is very poor.'

'I can't help worrying,' I admitted.

The evening dragged slowly and the programme on television about missing persons didn't help. Still no phone call. I tried to persuade Steve that he would like to go back into Plymouth for a drink. Eventually he couldn't stand it any longer and went. He called at her flat, but no-one was there. He phoned to tell me and my heart sank even further. Ted tried to comfort me, saying he thought she had probably gone off to the party in her usual rush and forgotten about everyone else who might worry.

'You can't do anything more tonight, it's too late.'

I sighed deeply and realized what he said was true as it was eleven o'clock. Ted managed to sleep. I didn't. I heard the first cock crow and for once I was glad to hear it. It was beginning to get light as I struggled down to the kitchen and made a cup of tea. At best it would be hours before I heard anything because if she had been to a party she wouldn't surface until midday. By Sunday lunchtime I was insisting that Ted or Steve went to the police. Then we had a phone call from Kate.

'Hello,' she said brightly. 'Sorry I didn't come over this weekend.'

'Kate,' I screamed down the phone, 'where the hell have you been? We've all been worrying like mad.'

'Have you?' She sounded surprised. 'I left a message with Vanessa to give Steve when she saw him.'

'Steve didn't see Vanessa,' I told her grimly. 'So we knew nothing.'

'Oh Lord (pause) I suppose you've been worrying?'

'Of course we have, you silly girl, we didn't know where you were or who you were with or anything.'

'I went to Jenny's. Her brother was on leave and she's keen on his friend so I said I'd make up a foursome. We stayed the night at Jenny's. I'm there now. I'm sorry.' Kate had her little girl's voice by now and what with the relief of knowing she was OK I felt myself softening.

'Thank goodness you're all right. Don't ever do that again, I hardly slept last night for worrying and imagining things.'

'Course I'm all right. I'm sorry you didn't get my message though.'

'So am I,' I said with feeling. 'I sent Steve into Plymouth to see if you were at the flat and no-one was there so—'

'Well, where was Vanessa, then?'

'I don't know where she was, and I didn't know where you were either.'

'Sorry, Mum, honestly. I thought Steve would see Vanessa.' The little girl voice again, sounding as though tears might be on the way.

'I was trying to get Steve to go to the police.'

'Oh, Mum,' Kate wailed and this time tears were arriving.

'Well, never mind, Kate, but if you ever do something like that again and don't let us know, direct, there'll be trouble, you hear?'

'Yes Mum. Sorry. Is Steve there?' I handed the phone over to her brother, thinking he might back me up, only to hear him ask 'Had a good time?' There was a long pause as Kate presumably told him that she had. My hands were shaking in relief and I didn't know whether to laugh or cry as I heard Steve's end of the conversation.

'Did you? Smashing. Yes, I bet you did. She didn't? Really? (chortle of mirth) Yes she was, worrying like blazes but she's all right now. What, tonight? Yes, I don't mind. What time? Right then, see you then. Bye.'

'Honestly, Ted,' I exploded. 'Kids are rotten. "She's all right now." How do they know I'm all right now? I'm suffering from shock and mental anguish.'

'Calm down, Liz, and drink this sherry. Can't you remember getting into this sort of trouble when you were young?'

'Well—' I began.

'Can't remember back that far?' queried Steve, grinning.

'Not only your sister but you are living dangerously,' I threatened. 'Wait till I starch your underpants or you need a fiver to see you through or—'

'Peace!' commanded Ted. 'I want to read the paper. The panic is over. You both have realized by now that your mother is a worrier, especially when it comes to her offspring, so don't do it again. Shut up all of you.'

'All of us?' I asked. 'There's only Steve and I here. The real culprit is missing.'

'Not any more, she isn't,' Steve grinned. 'I'm hungry, can we have dinner now?'

Steve has, in the last few years, taught himself to play the guitar. When he was little I suggested he might like to have piano lessons because he always liked music but he wasn't the least bit interested. When he announced that he had bought himself a second-hand guitar, we generously offered him the full use of our caravan. It's an old caravan, four-berth with electricity and gas laid on. Also, it's some fifty yards from the house. Every now and then I was allowed to witness the birth of a new chord and I admit he made surprisingly good progress. He took some lessons and tunes became recognizable. He exchanged his guitar for an electric one. Ted wondered whether to tow the caravan further away but one day he, too, recognized a Beatle tune and decided to leave things where they were.

One summer evening Kate and a girl-friend were staying with us and had asked if they could sleep in the caravan. Much scurrying back and forth with sleeping bags and torches, coffee jars and basic rations took place. Then they both took off to watch a local cricket match, or to be more accurate, to watch some local cricketers. Later, Kate poked her head round the door.

'All right if we ask some of the lads back for coffee tonight?'

It was about ten o'clock, just as a thunderstorm was well under way, when a crowd consisting of the whole cricket team drove up into the yard and Ted and I were fascinated to see a queue form at the caravan door and somehow find room there to get in. Steve returned with a friend in tow and, incredibly, they too squeezed in. We heard the guitar playing, roars of laughter and girlish giggles, above the sounds of thunder and lightning.

'Well,' said Ted, 'I don't know what you think but I say there's safety in numbers and I'm going on to bed.'

As we were reading in bed, much later, we heard sounds of people leaving. There were a few shouts and we heard Kate reassuring someone that it was only the goat. Ted remembered that he'd not shut Topper's door and we imagined someone's unsuspecting amazement if they'd encountered a goat in a thunderstorm in our yard. There was more laughter and eventual silence.

In the morning we heard of a highly successful evening and riotous reactions to Topper who had popped out of her shed in between storms to investigate the strange cars in the yard. Unfortunately, a door had been left ajar so Topper had got in the car and had a snooze. Upon being wakened and being told to get out she had shown her displeasure by depositing certain crumbs of disapproval. The owner of the car was going to have to live that down gradually.

Kate started to learn to drive. Her first attempts were in the Landrover with Ted but soon she'd mastered enough of the art to be let loose in our fields. She'd always been a competent rider and now she approached driving in the same manner. You could hear her whooping away as she bounced across the pasture.

'Whoa, Neddy, steady boy!'

'It won't take fences, Kate!' yelled Ted as he nimbly leapt out of her way.

Actually, on the road she was pretty good and I should know because it fell to Mum, good old Mum, to take her out in the white Mini we had then. Luckily we had a good spot to make for, the Water Board were building a new road that would eventually surround the new reservoir they were constructing and a car park had already been laid. This was where we did three-point turns and I must admit Kate's were better than I'd ever done prior to my test. Instead of her being timid about passing other drivers as I'd been, Kate was almost aggressive, and toward the end of a driving session I was in need of resuscitation. We would, therefore, often end a half-hour of 'Not so fast here, Kate'. 'It's all right, Mum, I know where the brake is'. 'Well, prove it then' and 'Mind, Kate, there's a car coming'. 'I saw it, I'll get him, don't worry' by coming to an abrupt skidding stop outside the Fox and Goose.

In a small community everyone knows what everyone

else is doing and we would be greeted with remarks like,

'I saw you two were out on the roads, that's why I didn't go any further than here.'

'You missed me just now, come in after me have you?'

'You're getting better, Kate, I see there's no grass on your bumpers.'

When she did pass her test and it was announced with some pride in the public bar, several locals threw their licences down saying they wouldn't dare use them any more.

Of course it was lovely for Kate to become independent and on four wheels, but it added to my worries. She always used to ride her tricycle in one direction and look the other and tended to be the same on horseback, and I didn't feel she'd ever change radically. She had quick reactions, I'll grant her that. She arrived one day, having driven Steve home. He came in dramatically holding his forehead. So dramatically that I knew it couldn't be serious.

'She's some bloody driver,' he moaned.

'Whatever's happened?' I asked. Kate followed him in, almost doubled up with laughter.

'I did an emergency stop,' she explained.

'Didn't you warn him?' I said.

'Not warn him, I didn't have any warning even,' Kate spluttered.

'Those lanes?' I hazarded. 'They're very tricky.'

'Not the lanes. It was a damn mouse. A mouse crossing the road, that's all, and Superdriver here slams on all anchors and we stop, dead. Only I didn't.'

'It'll be a good job when seat belts are compulsory,' I stated.

'Should have zebra crossings for mice,' Steve muttered.

Only very rarely do I venture into Plymouth to shop. This is partly because I don't like shopping and partly because I don't have the constitution for it. I usually drive in in the morning and face the hazards of trying to find a parking space. When I've achieved that I carefully lock the car doors and find I haven't the change for the dreaded machine in the multi-storey car park. I flap around begging from other parkers and each time I try hard to use the machine properly. Mostly it spews out a ticket for a different length of time from the one I was aiming for. I return to the car and have to unlock the door again to stick the ticket on the inside of the windscreen. By now I am exhausted mentally. I make for the nearest coffee shop to recover my spirits and try to remember what I came to Plymouth for. By the time I've found it, my feet are wailing miserably because they are more used to wellies than tidy shoes, so I go home. This is not the way my daughter manages me when she comes with me.

'Come on, Mum,' she hustles me off having expertly dealt with parking and ticket machines. 'We'll do this street first and then come down the other.'

'Couldn't we have a cup of coffee first?' I plead.

'We've only just got here,' she answers.

I try to keep up with her and obediently dive in after her to shops with music blaring. She gazes in rapture at some garment that looks in need of an iron.

'Isn't that absolutely gorgeous?' she mouths at me.

'What?' I shout. She repeats her remark.

'It doesn't look finished,' I say, holding out the bottom of the skirt which is unravelling all round.

'It's supposed to look like that,' I'm told.

In the next shop I see something I don't think is too bad.

'That's quite nice,' I venture.

'It's all right I suppose, for someone old.' I decide to keep my opinions to myself. Someone old, indeed, Kate's skirt is made of material that I can remember my grandmother liking and as for the shoes, well!

'Couldn't we stop for elevenses now, I'll pay,' I add as an incentive.

'We haven't done Marks and Spencers yet.'

'I want the loo,' I say in desperation.

'Honestly, Mother, you used to tell me off for wanting drinks and the loo when we were out.' I fall silent, she's quite right, I did. I must look crestfallen because I'm shoved toward a café where I collapse gratefully over coffee and a cigarette.

'You really shouldn't smoke, you know, it's awfully bad for you.'

'I know, but I eat if I stop smoking.'

'That's just an excuse.'

'I expect you're right,' I concede as I stub out the offending cigarette-end and wonder how long I have to wait for the next respite, when I can have another drag at the weed as Kate calls it.

'Now then, Mum, don't forget you wanted the loo.'

'All right Mummy,' I retort and we both laugh. 'Put it down to senile decay,' I tell her. 'But I need regular cups of coffee and cigarettes AND the loo especially when I'm in a strange environment.'

'Well,' says Kate, 'you should be OK for half-an-hour now, shouldn't you?'

'Cheeky little bitch,' I mutter and the old lady near us tuts at me, no doubt thinking that I set a very bad example. Which is probably true.

Chapter 13

It's Different in the Country

This is something my Mother often has reason to point out, either in mild surprise, puzzlement or wonder. Mother and many other friends who are city dwellers make comparisons: their traffic noise with our peace, their streets with our moors and more especially the slower tempo of our days in which we have time to take an interest in each other's lives. The story of our bed is a good example of this.

Ted and I haunted auction sales when we first came here in order to replace modern furniture with things that would fit in with our cottage. I had inherited an antique Welsh dresser, a court cupboard and an oak table so we tried to match up with them in spirit if not quality. We had enormous fun. In these parts auctions abound. The best ones were farm sales held in situ where the animals were sold first, so we had plenty of time to examine the household goods before the thrill of bidding began. The sheer unexpectedness of what turned up made me an addict.

However it was in a saleroom that our bed appeared, well, part of it, behind a huge wardrobe. I spotted the head and foot boards of mahogany. There didn't seem to be any more of it than that and I couldn't see a lot number on it. We both liked the look of what there was though, so I pursued a porter. He was elderly with grey curly hair and rosy cheeks. His bright blue eyes fixed on me as I asked him about the bed.

'That's not really in today's sale, Missus. You see, the rest of it has got lost, so we can't sell it as a bed.'

'What a shame,' I said. 'It looks in good condition, too, and it would be just what we wanted.'

'Well now,' the porter rubbed his slightly bristly chin. 'If you party are really interested in it, I could have a word with the auctioneer. You only need two bed irons and a set of springs to make it complete.' My heart sank. Where on earth do you set about getting bed irons, even if you know what they are?

'Now supposing you got they bits there, I reckon I can help you get hold of the irons,' suggested the porter.

'Could you? That would be lovely.' I was enthusiastic; Ted looked dubious.

'Ah well, I could help you. You party wait here a minute.'

Ted and I had a closer look at the bed bits. They were in good condition and certainly mahogany. Each upright piece was beautifully made and the centre one had a little flower design carved on it. The foot end was matching. The porter returned.

'The auctioneer says you make him an offer and you can have it,' he grinned. 'The truth is, it's been hanging around a while and he'll be glad of its space. Don't go making a silly offer,' he added.

'What d'you think?' enquired Ted. 'A tenner?'

'No need for that,' scoffed the porter. 'Offer him two pound fifty and I reckon he'll jump.' Ted went off to make his offer.

'It's very kind of you to bother,' I began.

'It's nice to see people who like the old styles,' I was told. 'I've never taken to these here modern deevans myself, too low down for me and no space under 'em. No damn use if you want a potty at hand, and I got an outside toilet.'

'Quite, I see your point,' I agreed. 'And can you help us to find the bed irons, I think you said?'

'I've got some at home.'

'Have you? And can we buy two from you?'

'Bless you, you can have two and welcome. I collect 'em.'

'Collect them? You must need them then.'

'Lord no, I got near on fifty of 'em.' I must have shown my puzzlement, so he explained, 'They make fine fencing posts. The old beds don't sell and we often end up getting them taken away for burning, so I keep the irons to fence my garden.' Ted came over to us smiling. 'You were right, I've got it for two fifty.'

'That's great,' I turned to the porter. 'And this gentleman is going to let us have the two irons.'

'You call me Bert,' he told us. 'Now, I'll tell you where I live and you can call in on your way home. Go round the back and you'll see plenty of what you want leaning against the wall. Mind you get two the same. And I've been thinking, you knock on next door, that's where my sister lives. I'm sure she has an old set of springs in her shed that'll fit you up.'

We were quite taken aback at his kindness, he really cared. He told us how to find his home and sure enough, as we walked round the back of a lovely old cottage we found a vegetable plot neatly fenced with old bed irons. They are the pieces that join the head and foot of a bed together. They just slot in, so as long as we chose two the same length, our problem was nearly solved.

'Well, we're in business now,' Ted said as he picked out two pieces from the collection leaning against the wall.

'I'm glad you finally persuaded him to accept some money, I'd have felt awful taking these for nothing like he wanted us to.'

'If I hadn't said to buy himself a drink with it I don't think he'd have taken it as plain cash.'

'He was a kind man who doesn't hold with modern deevans,' I explained. 'Shall I call on his sister?'

'Daren't not, he was very insistent.'

Bert's sister turned out to be as helpful as Bert.

'That's right, I do have a set of springs in the shed,' she agreed. 'And you party will be welcome to them. They're in good condition but I wish I'd known you were coming, I'd have cleaned them up a bit. I bet they're all dusty.' She was most concerned. However the shed turned out to be a weatherproof wooden construction containing deck chairs and other such items. The bedsprings were just the job. Once more we had a hard job to get her to accept anything for them. At last we arrived home triumphant with our bed. Although at the time we had no carpet on the floor we couldn't resist putting it together to see what it looked like. We were thrilled to bits, it looked stately and expensive. I spent the rest of the day polishing it. The day after we splashed out on a new mattress. That night, on bare boards and in a curtainless bedroom we used the bed for the first time. Because the light fitting was broken we stood our old oil lamp on a stool. It not only bathed the room in a soft glow, it took the chill off as well. We'd spent a hard day decorating and retired early, taking our sandwich supper with us. After settling down, Ted suddenly leapt out of bed.

'Got an idea,' he said over his shoulder as he went downstairs. Soon I heard him thumping back up again. To my utter amazement he came through the door backwards, struggling with our new dustbin.

'What on earth?' I spluttered.

'You'll see.' Ted set off again and returned with the portable television set. He upturned the dustbin and set the television upon it.

'There,' he said. 'Madam has telly in bed.' We sat up in state, no carpet, no curtains, just a super bed, an oil lamp and now television. We watched a programme on stately homes. I wasn't envious, we had all we could want. At least—

'Ted,' I giggled, 'there's only one more thing we need.'

'Oh Lord, what's that?'

'The loo is downstairs' I pointed out.

'You don't want to now, do you?'

'No, but I think we must get a proper commode to go

with this bed.'

'Huh, there's plenty of room for a guzzunder.'

'I know, guzzunder the bed. Much better than a modern deevan.'

Most of our visitors see our way of life in the summer. There are disadvantages to country living, of course, and getting from one place to another can be difficult. Until the family got to a self-drive stage, Ted and I were taxi drivers. We soon got to know other parents in the same position and it was possible to share journeys. One week we were on the taking end of getting teenagers to the local disco and the next week we had to stay up long past our bedtime to go and collect them at some ungodly hour. Still, my opposite number in town would have been worrying, no doubt, about offspring catching the last bus home with its possible dangers. In the country areas, collecting youngsters is the normal procedure so they don't suffer loss of face when Mum and Dad turn up. A car of some sort is less of a luxury than a necessity. Teenagers from farms have probably been driving tractors for years and passing the driving test is not much of a problem. Persuading them that an elderly car shouldn't be driven like a tractor *is* another problem. Teenagers have oldish cars as a rule and they're always going wrong. Begging lifts is commonplace. Occasionally, due to unforeseen circumstances, Kate is grounded. I once overheard a conversation between her and her brother.

'Steve, are you going out tonight?'

'Yes, why?'

'My car's at the garage.'

'Oh yes.'

'So I can't go out.'

'What about Jim's car?'

'Jim?'

'I thought you and he—?'

'No, that was off ages ago.'

'Oh.'

'Could I come out with you?'

'No.'

'Why not?'

'We don't know where we're going yet.'

'I don't mind where.'

'It might be somewhere unsuitable for you.'

'What d'you mean?' Kate gets indignant.

'I don't know yet, do I?'

'Who are you going with?'

'You wouldn't know them.'

'I might.'

'All right then, Tim and Alan.'

'I don't know them.'

'I said you wouldn't.'

'But I might have.'

'It makes no difference, you're not coming with us. I don't want my sister with me.'

'Why not?'

'Because I'll probably stay the night in Plymouth and crash out at Tim's.'

I realize at this point that I shouldn't have been listening. I'd been told that Steve was staying the night at a friend's. It sounded so much better than crashing out somewhere. Then Kate wanders disconsolately in to me.

'Fancy going to the cinema tonight?' I haven't seen a film in years.

'What's on?' I ask.

'Well, there's *Sexy Sylphs*, *Midnight Maniac at Large* or *What the Single Girl should Know*.'

'Er, no, I don't think so.' Kate retires huffily to her room from where I hear her transistor going full blast. Then Steve begins playing his guitar. A hair-dryer starts up. I wonder if the exposed beams and single floor-boards between upstairs and down is really such a selling feature. Ted and I can hardly hear the programme on TV.

'D'you know,' Ted says, 'I think we might just as well sample a bit of *Sexy Sylphs* or *Midnight Maniac*. Surely either would be more peaceful than this.'

The phone goes and complicated arrangements are made to get Kate to some party that's going on. Another phone call and Steve is gone. The house is silent, the dogs settle in front of the fire, all is peace. There's a sudden knock at the window. Dogs start barking, Ted struggles out of a sound slumber. I drop my knitting on the floor and answer the window. It's our neighbour, Joe Pascoe.

'I'm ever so sorry to bother you, Ted, but I got an awkward bugger of a heifer calving and Dad's out at Whist or he'd give me a hand.'

'That's all right, Joe, I'll be out. In the barn, is she?'

'Yes, I've got her in all right.'

'OK, I'll be there.'

The house settles once again, dogs doze, cats jostle for position in Ted's vacated chair. The phone goes. It's Kate.

'Mum?'

'Where are you?'

'At Mandy's house. We're having a party.'

'I can hear it.'

'That's the mobile disco, it's great.'

'It's loud.'

'No good otherwise. Look, can I bring Betty back to sleep? Only her parents are out and can't fetch her, she would have slept at Mandy's but there's already five sleeping here.'

'All right,' I tell her. 'I'll make up the spare bed.'

'Thanks Mum, and er?'

'Yes?'

'If Robin brings his sleeping-bag, could he sleep there too?'

'Who's Robin?'

'He's a super chap, Mandy's cousin's friend. They didn't know he was coming and there isn't room for him here either.'

'OK. If he doesn't mind the put-you-up?'

'Super, see you later. Don't wait up. Mandy's brother will run us home and he's ever such a good driver. He's got a TR7.' I gulp bravely.

'Have a good time,' I say weakly.

'I will, thanks Mum.'

In the morning I fall over a bundle sleeping on the front room floor. Why didn't he use the put-you-up I wonder, then realize that there is a bundle on there as well. I make a large pot of tea. The bundles start to move and heads appear and introduce themselves. The heads are licked awake by the dogs who are very pleased to welcome visitors. Sounds of giggling come from upstairs. The two male lodgers disappear in consternation to get their trousers on in the bathroom before the girls come down. I do scrambled egg for the four of them. Ted acquires two helpers to do the pigs with him. The girls help me. It turns out to be an enjoyable Sunday. I like teenagers, most of the time.

Chapter 14

The Quiet Life

I read somewhere that 'Good girls are the only people who can keep diaries because bad girls don't have the time' so, agreeing with that in principle, I found busy animal-keepers have much in common with bad girls. I enjoy letter-writing, when I have the time, but find that once confronted by a blank page I can't recall all those interesting things I wanted to say. I tried keeping an exercise book to hand in which I jotted down odd incidents and found it a great help. I came across one such notebook the other day. Much of it was missing and then I remembered what had happened to it.

One sunny afternoon I'd rebelled, ignored the ironing, the urgent need for a mending session, convinced myself that Ted and I didn't need to eat potatoes every day and taken myself into the garden. I'd catch up on my diary instead. I did a fair bit of recording before my eyes began to close. I was very comfortable lying in the sun despite Topper close to me, burping companionably and chewing her cud practically in my ear, she believed in close contact as did the collie who was panting down my other ear. At that time we had Topper's kids around, and as long as Topper was nearby I knew they were close to hand as well. I awoke with a start to hear the phone ringing and by the time I returned all my belongings had suffered from my absence. My sunglasses were under Topper and quite a different shape from the original, she'd eaten the rest of my cigarettes

(Topper was as addicted to tobacco as myself, except she chewed it) and the kids were happily nibbling at my exercise book. However, reading through the remains gives a fair picture of what made up 'nothing much' when friends asked what I'd been doing with myself.

This Easter I tried something I've been meaning to do for ages. I peeled an onion and wrapped the brown skin around an egg. To keep the skin in place I put the wrapped-up egg in one of those plastic nets you buy nuts in. Then I boiled it all for about twenty minutes, fished the rather smelly object out and left it to cool. When it was unwrapped I was most satisfied to find a beautifully mottled egg of intriguing design. I presented Ted with one on Easter morning.

'I don't have to eat it, do I?' he asked, his brow wrinkling.

'No, of course not. It's to look at,' I reassured him. 'I made it for you.'

He gingerly placed it on the window-sill where it has remained.

This April has been sunny and pleasant. Kate's already hoping to get a sun tan so she decided to hurry things up by investing in a small sun-lamp. I, too, had dreams of getting bronzed until she took it to her flat so that Vanessa, her flatmate could use it as well. However, she left the box behind and I've been very pleased with it as a hospital for a duckling.

I'd given the responsible job of hatching a couple of duck eggs to a broody hen but she'd only had a fifty per cent success rate, resulting in a crazy mixed-up duckling that I can only call Charlie. I spent this morning dashing out on rescue missions; first he got stuck between some chicken-wire fencing and a grassy bank, then he fell into an empty enamel bowl that he couldn't get out of and finally tried to join the wrong family group and received a nasty peck on the head for his impudence.

I've administered first aid by way of antiseptic cream with a small square of gauze on his head, and popped him into the sun-lamp box looking like a bald-headed man on holiday with a knotted handkerchief to keep the sun off. I think he'll recover as he is pecking happily at some chopped-up hard-boiled eggs. I hope Kate doesn't want the box back as Charlie's droppings are liberal and smelly.

April 21st
I am suffering from kids. Topper has had her babies and they are all male. They are absolutely gorgeous to look at and hell to live with. Topper is a super mother and is feeding them well so they are full of energy and have the confidence to go anywhere. At first, Topper would suckle them and then park them somewhere she thought suitable (from where they wouldn't move an inch until she returned), but her ideas of suitability and mine didn't always coincide. Yesterday she parked them in the outside lavatory, one either side of the loo and one on the lid which was luckily down. The day before, she found the front door ajar and tucked all three of them behind it. They moved around to get comfy and closed the door, so there they were stuck between the front door and the porch. At first all was well but when Topper came to collect her brood she couldn't get at them. She was calling them and they were yelling blue murder because they couldn't get to Mother's milk bar. I heard the rumpus over the noise of the vacuum cleaner and had to go and organize the reunion.

Today is the worst incident so far. Topper took them along to our neighbours garden. Why, I can't imagine. We are surrounded by our own land and open moor but she had to go to their garden. The result was dreadful, what they haven't nibbled they've trodden on. Ted is even now fixing up a means of tethering Topper; if she's under control the kids won't stray without her. I just hope we can come to a happy arrangement.

April 23rd
Thank goodness, Topper doesn't mind being tethered. Mind you, she has every amenity to hand: a portable shelter in case of rain, wind or too much sun and a special bucket of water that stands in an old tyre so it can't get upset. We shall move her every day and we have to give her extra feed to make up for lost grazing rights, but life is easier.

April 25th
The hens were sunbathing outside the barn today. One was lying with her wing outstretched in the sort of position that used to get me panicking about dying poultry, now I'm wiser and know they get drunk on sunshine. As I walked past she opened one eye to see what was blocking the sun, then closed it again when she saw it was only me; after all, we have known each other since she was an egg. One young pullet was pecking at the grass when a handsome cockerel strutted up. She pretended not to notice and continued pecking. He gave a fruity growl and came nearer. She turned away, playing hard to get but knowing full well that by showing him her feathery petticoats she was really giving him the come-hither. There was a flurry of activity during which he had his wicked way with her and he stalked off. She shook herself thoroughly and, purely for the other hens who might have been watching, clucked a bit indignantly, but you could see from the gleam in her eye that she thought herself quite a gal.

The sun was so lovely that I decided to give the geese a treat. I filled an old bath with water. I think they heard the splashing because they appeared across the lane like a flotilla of galleons. At the same time a car came along. Geese and vehicle met in the middle. I made myself scarce, there is nothing they like better than holding up traffic. The driver stopped. Once the car was stationary the geese spread themselves out and did a bit of plumage rearranging. The driver revved his engine, hoping, no doubt, to intimidate them into moving. They ignored him. He used the horn. This was a signal for them all to reply. They closed ranks and stretched out their necks at the car's bonnet. Having honked noisily at him and made their point they remembered they were heading for water, so they waddled slowly and with great dignity off the road and into the yard. The car moved on, I came out of hiding and bath-time began. The farmyard pecking order governs bath-time as well as feeding, so the gander had first splash and his most junior wife had to make do with a lick and a promise in very little water full of debris from the others.

April 28th
Charlie the duckling made a complete recovery but never took to his foster mother. I've given him to a friend who has other ducks and he's settled well. Boris the Muscovy drake had totally ignored Charlie so I reckoned it was time Boris met a good lady duck. My friend hadn't the right make of duck to help so I wrote a notice for the local post office window. Ted was appalled at what I composed and swore he wasn't going to answer the phone but determined on a quick result, I wanted my advert to attract the maximum attention:

LONELY MUSCOVY DRAKE, HANDSOME BUT FRUSTRATED, WISHES TO
MEET A MUSCOVY DUCK WITH VIEW TO MATRIMONY. WHEN PHONING
PLEASE ASK FOR BORIS.

It worked. Today I collected a duck named Matilda from a local chicken-run where she, too, was lonely. I introduced them to each other in the privacy of the big barn. As soon as the two blind dates met they fanned out their tail feathers and Boris sort of hopped from one foot to the other. I could see the engagement would be short, in fact consummation seemed imminent so I closed the barn door gently.

April 29th
I have brought my cup of coffee outside and I'm sitting on the bank watching a beetle. It climbed up a tall blade of grass and I was wondering what it would do at the top. The sun glinted on its back and I couldn't decide if the beetle was green or gold. The grass stalk began to bend under its weight and just then it produced a pair of black tissue paper wings and flew off. While I was engrossed in nature, Peggy the hen had her head in my cup of coffee. She is drinking from it daintily, tipping her head back to swallow, keeping one golden eye firmly on me as much as to say 'Share and share alike, you have my eggs, I have your coffee.' Now she's shaken her head and there are coffee splashes all over this page.

April 30th
Boris and Matilda are honeymooning, completely engrossed with each other, so I do hope the outcome will be ducklings this year.

The car hasn't been well, refusing point-blank to start unless on a slope. I daren't use it for fear of being stuck somewhere. Ted can cope, he has a wonderful way with it. He opens up the bonnet and appears to beat hell out of it with a lump hammer and a piece of wood. Sadly, I haven't his mechanical skill so I shall take the dogs up on the moor.

Had a lovely wander over Brown Gelly and on the way home saw something white flapping on the ground. Thought it was a plastic bag but discovered it was a seagull with a broken wing. Couldn't bring myself to leave it there so picked it up. I had to throw my anorak over it to avoid a vicious beak. Ted saw me coming with the bundle clutched to me. I was quite breathless, having found that a seagull is no lightweight, and the dogs had been jumping up at it as well to know what I had.

'Not something else to look after,' were his words when I'd unwrapped it.

'I couldn't just leave it for a fox, could I?'

'Why not?' said Ted with feeling as the gull attacked his thumb. We managed to use sticky tape and fix the wing in position. The result was one set wing to us and several bitten fingers to the gull. I've left it in the outside loo with a bowl of water and a herring in it from the deep freeze.

May 1st

The gull is doing well on a diet of raw fish, bacon rind, cheese and my fingers. I feed him several times a day in the outside loo. It's a messy business. I wrap an old towel around me and have him on my lap, then I tap on his beak which seems to encourage him to open it, at which point I try to pop in a bit of food but nine times out of ten he gets bits of my skin as well. I embarrassed the postman today, I was sitting on the loo feeding the gull as described, with the door open, when the post came. I called out 'Good Morning' as usual but realized from the look on his face that he couldn't see the gull.

One of our grown-up calves has come into season and she's blaring to the world in general that she needs a bull and quick. Joe Pascoe said to me last evening, 'I see one of your heifers is rummaging.' I must have looked puzzled because he laughed and explained that rummaging means calling for a bull and he went on, 'If she don't get what she wants she'll be off through fences and over banks until she finds a bull. You tell Ted to come and get Ferdy, he'll be glad to oblige her.'

So this morning Ted went off to escort Ferdy to our field. The big bull lumbered steadily along the road with Ted in the rear. I was stationed at our gate to let him in. The heifer smelt him coming and was leering over the hedge fluttering her long eyelashes. Ferdy got the message and quickened his gait. I was most relieved when he turned smartly into our field, not giving me a second glance. That's the way I like bulls, completely disinterested in me.

May 3rd

When Jess the collie had her seven pups on Christmas Day we knew that when the time came we'd have our work cut out to find good homes for them all. Jess was a super mum for the first fortnight and then, considering her duty done, she lost all interest in them. She lay with them only under sufferance and, wearing a martyred expression, gave them all of five minutes sucking time before standing up and shaking off the pups in assorted directions. Somehow I kept her at it for a bit longer and the pups thrived. I became their main object of affection. I was the source of extra rations, I gave cuddles and played games and I was the one who gave them an old blanket rolled up and tied with string to snuggle up to at night when Jess finally refused point-blank to go near them. Luckily we got homes for all but one. Ted is convinced that it was a put-up job from the start because the one we were left with was the smallest and my favourite. When prospective buyers came he would huddle near my feet and, yes, I aided and abetted him by pointing out the good things about the others. So we were left with Timmy. I called him Tim to start with, a good short name suitable for a sensible dog. The fact that he shortly became known as Timmy Muffin shows a change of plan. Timmy was strong in affection, enthusiasm and energy and a howling success as a clown but not too well endowed with grey matter. I have been doing a bit of training with him but as soon as he has an audience he goes to pieces. Take today's effort for an example.

'Ted, do watch Timmy a minute. He'll stay now, beautifully.' Ted looked dubious but watched dutifully.

'Timmy,' I said in my best Woodhouse manner, 'sit'.

Timmy collapsed at my feet gazing at me in adoration, tongue dangling.

'Now— stay,' I commanded. Timmy sighed audibly but stayed as I began to walk away a few yards. I was just beginning to feel a bit smug when I was thumped in the back of the knees by a bundle of black and white fur who thought I'd left him behind accidentally. I buckled on to the wet grass and heard Ted guffawing as Tim, with both paws on my shoulders, licked my face.

'Jolly good,' enthused Ted. 'You are getting on well.'

May 6th

Matilda and Boris have done it. We have duck eggs! We also have goose eggs and the gander marches up and down outside the goose house as if he's on sentry duty while the laying takes place in the morning. We all give that area a wide berth, the gander means business. The size of enemy does not deter him and he attacks everything: passing chickens, dogs and, this morning, the big lorry bringing the pig feed. His beak sounded like a machine-gun as he pecked at the wheels and the big burly driver refused to get down from his cab until Ted shooed the gander back to the goosery.

May 25th

Matilda is sitting on her eggs. Boris hung round for a bit but got increasingly bored by her inactivity. He's gone back to his bachelor habits and spends his time once more under the kitchen window and dozing on the doorstep.

May 26th

Today is wet and windy. I had a letter from Stella and lost the dishcloth. Stella and I met at training college from where she progressed successfully through primary schools to marriage with a financially clever businessman and continued with her teaching career. In her letter I read that she'd changed her car again, redecorated the lounge because she'd got tired of the curtains – 'and you can't change the curtains without altering the decor, can you?' Also she was about to be appointed head of her school so she was relieved she had found a reliable cleaning woman. I gazed round at my curtains, wondering if there was a decor that could match their faded state, and noticed the dirty windows where the cats rubbed their heads as they mewed pitifully to be let in. Stella, I recalled, didn't approve of pets. Then Buttons decided to climb up on my lap by way of mountaineering up my leg. He used his claws as cramp-irons and I slopped my coffee so I reached for the dishcloth. It wasn't in its usual place, nor under the washing-up bowl nor yet lurking behind the soap powder packet. I sighed and squinted down the side of the Rayburn. There is a three-inch gap between the wall and the cooker where I have found many a strange object, i.e., one sock drying (only one got wet, huh?) the dog's latest bone and shrivelled-up peas (I admit I know how they got there) and on one occasion I found a pair of Ted's underpants (my mother had mistaken them for the floorcloth, an understandable error, and had used them as such and left them to dry over the Rayburn from where they'd slipped into the dreaded gap). I bet Stella's Mrs Thingy would never allow such things to happen. I tried to get on dishclothless but I was unnerved. Ted came in for a tea-break and sat down.

'A cuppa, Liz, I'm spitting feathers.' This was a new expression he'd just picked up from Joe. Then a puzzled look came over his face. He stood up and peeled the missing dishcloth from his rear.

'What on earth . . . ?' he began.

'Sorry,' I said, 'it's the new daily – terribly forgetful.'

May 30th

Matilda has three ducklings. They are charming and Boris has inspected them and approved. He was all for joining his new family for a stroll but Matilda insisted he walk at the back so he retired huffily.

Ted brought up a middling-sized pig from the weaner house. It wasn't actually ill, I think it had been bullied so I set about spoiling it in the barn, offering Topper's milk and some beaten egg as an incentive to recovery. I think it'll do all right, its eyes are bright and that's a good sign.

June 2nd
I've been asked to be in charge of the Welly Throwing at the village fete. I think this came about because I mentioned to someone that we seem to have a collection of left Wellington boots with no rights. Ted has made me a notice board to stick in the ground with a sign on it for the occasion. As the back of it looked bare I have composed a history of the sport:

The History of Welly Throwing

A hungry farmer came in from harvest for his dinner. He was told there were pasties and he was so excited that he threw one of his boots in the air for joy. This became a habit and he found it improved his appetite so he recommended the pastime to his friends and so the sport began. Nowadays the word 'welly' is commonplace:

'Welly do love a good pasty.'
'Welly didn't make much of a job of that.'
'Wellies probably down the pub.'

Ted read it through, gazed at me thoughtfully, put it down and went back to his paper.

'Welly ain't very interested,' I muttered to Buttons.

June 4th
The convalescent pig has made great progress. This morning it progressed from the barn to the outside room where it forced entry somehow and rooted amongst the buckets and

wellingtons before finding a basket of vegetables. By the time I found it it had eaten most of the carrots, the best half of a cabbage, some onions and part of the basket. I convinced Ted that it was quite well enough to join its companions in the weaner house but whether it's due to my nursing, Topper's milk, the vegetables or the beneficial qualities of wickerwork, I'm not certain.

June 23rd
The seagull has gone. The sticky tape keeping his wing in position did a fine job until Joey, as I called him, discovered he could tear strips off. By the time his wing stayed in the right position by itself, not hanging down as it had when I'd found him. I let him out into the vegetable garden where he had plenty of space, but with wire netting round him so he couldn't make a premature break of freedom. At first he made no effort to use his wing at all, but gradually I saw him giving test flapping demonstrations which soon had more and more strength in them. This warming-up led to running hops and loping progressions which lifted him off the ground for a yard or two. His broken wing wasn't as good as his other and Ted remarked that he'd probably be the only seagull around that flew in circles getting nowhere. One day he made a really determined effort at flight but misjudged the height of the fence and crashed into it, landing unharmed. I took him to the open field the next day where he flew quite well but to my amazement he always came back to his starting-point. Later in the day some other seagulls flew over and whether he knew them or whether they merely encouraged him I don't know, but he flew up into the sky and disappeared into the distance without a word of goodbye. I wished him well and turned to the unpleasant task of scrubbing out his erstwhile home which stank of fish.

June 25th: Harvest Time
Today Ted and I were giving a hand with haymaking. It was hot and humid and no way could anyone take it slowly as thunderstorms were obviously in the area and Joe Pascoe had two fields cut, ready to bale and store. As often happens in country parts, unexpected help arrived. Two young chaps who were staying in the village thought they'd holidayed enough and needed some exercise, a fact they'd mentioned to Sid, the landlord of the Fox and Goose, who had promptly despatched them to Joe's hayfields. It turned out they'd both done haymaking before, which was a relief, because often help is offered and then withdrawn hastily as people find out what heavy work it is.

However, these two proved their weight in gold and stayed around a couple of days. Their reward came in two parts. The first one was the enormous harvest dinners and suppers the Pascoes insisted that any helper had, and, secondly, two attractive cyclists who asked if they might camp on our field for two nights. The girls wandered along to the scene of activity and to give them their due they helped as much as they could – they also did tons toward raising the flagging energies of the men, Joe and Ted included.

The harvest ended in a thunderstorm but with all the hay safely gathered in and a promising-looking friendship blooming between the two young men and the cyclists. As the first drops of rain fell we all gathered in the Pascoes' kitchen where, replete with home-made pasties, pies, cold meat, salads and cans of beer we sat about with our feet stretched out before us, trying to flex aching muscles. Bits of hay irritated us in places where you wouldn't think hay could get, and scratches on arms and legs were compared.

'Get your squeeze-box out, Meg,' suggested Joe. Meg complied and one of the young men produced a mouth organ. A good old-fashioned sing-song resulted which lasted throughout the thunderstorm. The girls ended up sleeping in our house and the two young chaps gallantly rescued their camping equipment and slept at the Pascoes. As far as we know the four of them continued their holiday together.

July 2nd
I have decided to do Cream Teas. So many visitors stop at our cottage to ask where Dozmary Pool is, and having explained it's just half a mile on I am then asked where they can get a cup of tea. No-one else provides teas so I thought I would have to go. Ted was a bit wary at first but once I explained that he wasn't involved other than to make a sign, he was enthusiastic. He put up a sign for me which Topper finds very handy to rub against and which both dogs water each morning, and I have made a batch of scones. I have clotted cream at the ready and home-made jam. Now, I am waiting to see if anyone calls.

July 4th
Am flushed with success. I had twelve visitors the first afternoon and eight yesterday. They all enjoyed themselves and were anxious to hear about life on the moor. The only problem is keeping Topper out. Her kids have gone now, and we have been letting her free range again but she is so intrigued with the extra population that I have had to compromise. She is tethered near the front door and is proving an added attraction. One brave party of four decided to eat outside in the sun but I am sure they gave most of their tea to either Topper, who loves scones, or Peggy the hen, who likes anything (I caught a glimpse of her on the table at one point). Still, as the customers seem to like the wild life as much as their tea, who am I to shoo away the animals?

This is where my diary ends or rather where it has been eaten. I didn't keep one at all during the cream tea season. I did enjoy my catering experiment, and will probably repeat it next year.

Life has been described as a rich tapestry. My life is certainly varied but I am aware of the knots at the back of it all.

Chapter 15

Snow

Mother arrived at Christmas. Ted had fetched her and in his absence I'd cleaned and polished the cottage. I'd even had time to place of bowl of holly with a scarlet candle in the middle on the dining-table. I made a pot of tea as soon as Mother arrived and we were drinking it when I heard some strange sounds from the front room. In the confusion of bringing Mother's luggage in the front door had been left ajar. Now I found Topper with her forelegs on the table helping herself to the candle. She had it in her mouth like a fat cigar and the holly arrangement was now a collection of stalks.

'Good job we hadn't put the Christmas tree in there,' said Ted.

'She was probably on her way to my bed,' Mother said dryly.

We told Mother all the local news and Kate gleefully announced that Jess the collie was due to have pups any minute now.

'I trust she isn't going to have them on someone's bed.'

'Of course not, Mother,' I assured her. 'She sleeps in the kitchen.'

'Well, I sincerely hope we get Christmas dinner out of the way, first.'

Early on Christmas morning, Mother, who is an early riser, was heading for the kitchen to make her first cup of tea when she heard squeaks.

'Liz,' she called to me, 'there's new sounds coming from the kitchen, is it all right for me to go in?' I hurried down and we went in together. Jess was in her bed surrounded by a squirming heap of black and white puppies. She looked very perplexed and anxious. She always looked guilty and seemed to assume blame. She realized these pups were her fault but she didn't know what to do about it. I praised her and patted her head and made a great fuss. Jess relaxed; perhaps these new babies weren't going to get her into trouble after all, perhaps she could get to like them if they brought all this praise and attention.

'Clever girl, Jess, how many have you got?'

'Five, isn't it?' said Mother.

'Six, I think, no, seven. Jess, you've got seven babies!' By now the whole family was in the kitchen.

'What a super way to start Christmas Day,' I gushed.

'Timing was never your strong point,' Mother stated. 'But they are rather sweet. What will you do with seven of them?'

'Keep them all,' suggested Kate just to aggravate her.

'Oh, you couldn't keep all of them!' Mother was flustered.

'Just to start with,' I calmed her down.

'They'll be in the way getting the dinner,' she warned.

'I'll move them.'

'Oh good.' Mother sounded relieved. 'Where will you put them?'

'I'll move them to the other end of the kitchen.'

Mother tut-tutted a bit under her breath but she soon got used to the nursery in the corner.

There was a snow warning soon after Christmas. Ted and I weren't too worried, the pig feed was delivered in bulk and our feed hoppers were full. The deep freeze was well stocked in case of adverse weather conditions because I knew from the Pascoes that being snowed in wasn't impossible. In any case, as I pointed out to Mother, there wasn't anything you could do about it. I am sure she felt that a strategic withdrawal to a hotel miles off the moor would be her way of dealing with the situation, but she listened to the weather forecast stoically.

'I loathe and detest snow,' she said. 'I always have, but at least here I won't be stuck on my own like I would be if I was at home in the flat. Good job you did that shopping yesterday, Liz.'

'That was just some essentials,' I pointed out. 'Like cigarettes and your gin.'

'And very glad I am that I remembered it.' She sounded quite smug.

'We shouldn't be short of food,' Ted told her. 'Liz stocked up months ago when Mrs Pascoe mentioned it. We lived like kings for a bit.'

'That was in November, I had to stock up again since then and not tell Ted about it.'

'There's a cupboard full of pet food, I know that,' Ted groused. 'But I'm not allowed to go to the deep freeze.'

'That'll be nice then,' Mother sounded a bit sceptical. 'I know the pig feed came and now I know the pets are all right and although I'm always pleased to have a gin and tonic, I do like a little food now and then.'

'Don't worry, it probably won't snow at all.'

After tea, Ted looked out of the window and announced that it was sleeting. When I went to let the dogs out later on they recoiled in horror at the door. There was a white curtain coming down at a dizzying pace. The ground was already under about an inch of snow. I firmly pushed both dogs out to do whatever a dog's gotta do and they returned hurriedly, shaking snow off their backs and all over me. As we made our way to bed, Ted switched the outside light on and we could see the yard, an area of virgin white sullied only by dog tracks which were rapidly being obliterated. Lying in bed that night I was aware that a wind was rising.

In the morning there was a blizzard and the sound of outdoors came straight from a track of *Scott of the Antarctic.*

We couldn't see anything for the whirling curtain. I knew now what it felt like to live in one of those glass globes that you shake up to create a snowstorm. Ted had to get to the pigs and dressed accordingly in oilskins, a bit of baler twine round his waist to keep the wind out and a fetching little pink woolly hat that belonged to me. As he opened the back door the two dogs cautiously crept after him. A blast of wind and flakes invaded the outside room and I had a hard job to close the door behind them.

'Where's Ted?' Mother demanded when she came downstairs.

'Gone to feed the pigs.'

'He can't go out in this?' My town-bred mother was horrified. 'You can't see a thing.'

'Well, they must be fed,' I said. 'The Pascoes will have to get out on the moor to feed their cattle. Farming can't stop for the weather.'

An eskimo came to the window, white from head to foot and sort of leaning into the wind.

'Just come back for a shovel, it's drifted against the doors down there, I'll have to dig my way in. The shovel's in Topper's shed. Hand me her bits and I'll take them to her.' He had to yell for me to hear him. I collected Topper's 'bits'. Some dry toast, a couple of cracker-biscuits, some vegetable waste and, nicely disguised underneath, some proper goat food. I handed this to Ted, the dogs shot back in and I closed the door. About five seconds and I was plastered with snowflakes. I brushed them off before they could melt and turned my attention to the dogs. Jess shook herself as only a long-haired dog can. Her feet stayed still and everything else swirled about a lot. Moisture flew and suddenly she was almost dry but Potter, my old sealyham, stood disconsolately in his woolly fur that was encrusted with clinging snow. He'd never managed to shake himself, ever, and now as I gazed at him, slow drips fell off his undercarriage into puddles at his feet. I'd had trouble with him in this sort of state before. I found some old towels put aside for dogs but first I brushed off the worst with Topper's grooming brush. Then I rubbed vigorously. Jess joined in the game and tried to pull the towels away from me. Finally, to their disgust, I left them to 'finish off' in the outside room. Thank goodness for such a place, what would one do without it?

Partly because I wanted to go out and see what it was like and partly because I had things to do outside, I dressed suitably and emerged into a strange world. The wind was almost gale force and now the snow was coming down horizontally, as fine-grained stuff that stung my eyes and found its way down my neck and up my sleeves. I visited Topper first and renewed her water. We had sold her kids so I had to milk her. The bucket had been nearly blown out of my hand coming across the yard and I arrived with a crash as I banged the door shut behind me. Topper looked a bit startled. She'd not seen me dressed like this before and she quickly set about the fringe of my scarf. She nibbled, I milked and was slowly strangled in the process. Peggy was in there with her and they had both summed up the weather conditions and seemed content to stay in. I gave Topper a hay ration and left them. The milk partly blew out of the bucket as I crossed the yard again.

My next stop was the big barn. All the poultry were there, snug and warm. I made sure they had water and scattered their feed on the floor amongst the loose hay so that they had something to do looking for it. They were chattering happily enough together and I made sure the doors were shut; no point in letting them out today.

Our cattle, all four of them, were contentedly chewing at their hay. They turned steady gazes toward me and I could guess they were summing up my appearance. 'Suppose that's her winter coat, she'll look better when she loses it and gets a bit of sun on her. Mind you, I never thought she was a good doer, not enough meat on her for that, rushes about too much.' I went away from their shed feeling glad we hadn't animals out on the moor to worry about. Ours were all snug. I popped in to the pigs. Their houses, being insulated, were warm as toast and Ted had shed his layers as he fed them. The piglets under their infra-red lamps looked like holiday-makers at the beach. For two pins I'd have got under one with them. Ted and I battled our way back to the house for breakfast. It took us ages to shake snow off our clothes and find enough nails to hang it all up to drip. We were just ready to nip into the kitchen without the damp dogs when I had a thought.

'Ted, have you seen the geese? I've been to everything else.'

'Yes, they're out. I opened their door to make sure they had water and they rushed me.'

'Will they be all right?'

'They seemed to be enjoying it. I've left their house open a bit so they can go back in when they like but they were having a gay old time, flapping their wings and honking.'

'Oh well, they must do what they like, I'm not putting all that gear back on again.'

We had a huge breakfast and watched the weather change gradually from a frenzied blizzard to a steady snowfall. It stopped shortly before dusk as Ted was feeding the pigs again. I ventured forth to encourage the dogs as much as anything. Jess loved it, she rolled and she kicked and she barked. Potter eyed her balefully, I could see him thinking 'Stupid female, got no sense at all' and then, as she flashed past splattering us both with snow, 'Thoughtless, that's what she is, typical of the younger generation', and he made his way cautiously to his favourite watering-place only to find it under a mini-drift. He performed against the gate post, then gave a few stiff-legged kicks which resulted in a lump of snow landing on his back and haughtily made for the door. I peered along the road. There was a six-foot drift completely blocking the way to the Pascoes. The road was clear outside our gate for about two hundred yards and there was a another drift further along blocking the road in that direction. We were cut off!

The wind came back that night and sculpted the drifts into modern art shapes. I felt guilty because I knew the havoc the weather was causing all over the country and I was enjoying myself. Topper and I investigated the possibilities of snowsports; the geese went skiing and Jess and I ploughed into drifts just for the hell of it, Topper was more cautious but wouldn't go indoors, not while something was going on anyway, and as I decided to make a snowman just outside the gate she realized that a supervisor was needed. She watched my every move and was fascinated as I added an old scarf and cap to my sculpture. Ted was embarrassed at my frolics.

'Whatever will the snowplough driver think of that?' he demanded. 'Snow's no joke to him.'

Actually when the driver did arrive he thought it great fun and over a cup of tea told me he hadn't seen a snowman on the moor for many a year.

'Tell you what, Missus,' he said as he left. 'I think you should let your old man in now, reckon he's been out long enough,' and he clambered back into his vehicle chuckling to himself.

Mother rang a friend to find out how the city dwellers were getting on in the snowy conditions. It appeared that her friend wasn't enamoured at all and I could hear Mother commiserating.

'I know, it must be awful there and nobody loathes snow like I do, but it really has looked so beautiful down here and, well, it's different in the country, isn't it?'

THE END

A Selected List of Fine Non-Fiction Titles Available from Corgi Books

WHILE EVERY EFFORT IS MADE TO KEEP PRICES LOW, IT IS SOMETIMES NECESSARY TO INCREASE PRICES AT SHORT NOTICE. CORGI BOOKS RESERVE THE RIGHT TO SHOW AND CHARGE NEW RETAIL PRICES ON COVERS WHICH MAY DIFFER FROM THOSE ADVERTISED IN THE TEXT OR ELSEWHERE.

THE PRICES SHOWN BELOW WERE CORRECT AT THE TIME OF GOING TO PRESS.

□	99065 5	The Past is Myself	**Christabel Bielenberg**	£2.95
□	09373 4	Our Kate	**Catherine Cookson**	£1.75
□	12553 9	Swings and Roundabouts	**Angela Douglas**	£2.50
□	12033 2	Diary of a Medical Nobody	**Dr Kenneth Lane**	£1.75
□	12465 6	West Country Doctor	**Dr Kenneth Lane**	£1.95
□	12452 4	Out on a Limb	**Shirley MacLaine**	£1.95
□	23662 8	Don't Fall Off The Mountain	**Shirley MacLaine**	£1.95
□	12399 4	Any Fool Can Be A Pig Farmer	**James Robertson**	£1.75
□	12577 6	Place of Stones	**Ruth Janette Ruck**	£1.95
□	12513 X	Suffer Little Children	**Elizabeth West**	£1.75
□	12072 3	Kitchen in the Hills	**Elizabeth West**	£1.50
□	12707 2	Garden in the Hills	**Elizabeth West**	£1.25
□	10907 X	Hovel in the Hills	**Elizabeth West**	£1.50

All these books are available at your bookshop or newsagent, or can be ordered direct from the publisher. Just tick the titles you want and fill in the form below.

CORGI BOOKS, Cash Sales Department, P.O. Box 11, Falmouth, Cornwall.

Please send cheque or postal order, no currency.

Please allow cost of book(s) plus the following for postage and packing:

U.K. CUSTOMERS—Allow 55p for the first book, 22p for the second book and 14p for each additional book ordered, to a maximum charge of £1.75.

B.F.P.O. and Eire—Allow 55p for the first book, 22p for the second book plus 14p per copy for the next seven books, thereafter 8p per book.

Overseas Customers—Allow £1.00 for the first book and 25p per copy for each additional book.

NAME (Block Letters) ....

ADDRESS ....

....